AF327047
daniel rounds
notes on the possession
of animals by spirits

notes on the possession of animals by spirits

daniel rounds

Ad Lumen Press | Sacramento | 2023

Ad Lumen Press
American River College
4700 College Oak Drive
Sacramento, CA 95841

Part of the Los Rios Community College District

First Printing

notes on the possession
of animals by spirits

•

[PROLOGOMENA]

[FIRST PART:
ON THE EMERGENCE
OF MEANING FROM MATTER]

[SECOND PART:
ON THE PERFORMANCE
OF MEANING BY PEOPLE]

notes on the possession
of animals by spirits

for Daniel and Georgia

[PROLEGOMENA]

[sketch for an installation]

A white room in a museum. A stone pillar in the center of the room. Symbols cover the pillar in a dark veil of carved asemic script.

Those who pass time in the room notice that the lighting simulates the overhead movement of the Sun. Light emerges in the room's morning-east and recedes, hours later, in the retreating evening-west so that the changing light of the room plays with the stone pillar's shadow.

When night comes the room dims, and the edges of the megalithic stone pillar blur even as the lights of the night sky appear on the room's walls and ceiling.

Slowly, the room comes alive. The Moon, stars, and Milky Way move across the walls and ceiling, flickering as they migrate over and along the same paths they would travel in the night sky outside the building.

The installation in the white room can be visited any time. The stone pillar is always present. The room is always open.

[a dark room and two voices]

Another room, this one dark. The room has a stage and two players in black wearing Japanese Kyōgen Saru masks are positioned on the stage. Player one is seated on a large stone stage right. Player two is perched in a tree stage left. The two players gesture as they speak.

player one: how?

player two: what do you mean, how?

player one: I mean: how? as in, "how will it be told?"

player two: well, there's really no way to tell it.

player one: what do you mean?

player two: I'm saying there isn't any way to say it.

player one: so the story can't be told?

player two: no. I'm not saying that.

player one: but the story must be told. a reason must be given...

player two: no, not at all, but if you choose to tell the story, you could do so a hundred different ways.

player one: wait...do you mean ...

player two: do I mean what?

player one: what do you mean?

voice two: I mean nothing.

The room goes dark.

[three scenes for a theatre of one]

(intended for film)

[scene for a theatre of one]

an empty landscape, minimally descriptive yet expansive in scale.

imagine an extended lateral backdrop reaching both left and right as if an enormous roll of paper had been stretched across the reaches of the infinite.

and now volumes and recesses of empty space so that there is an immensity of depth, stretching forward and backward, as if one were looking into an endless vault of nothing.

and time. time as an eternal present experienced moment to moment, as now, and now, and now.

and now a figure appearing in the backdrop, a small figure, emerging and moving both over and against distance.

the figure moves inside the total horizon bisecting the amorphous extended reaches of all of this everywhere. the figure advances in shadow, a silhouette of smudge and blur, plodding forward, from horizon and backdrop, down screen toward a position closer to the reader.

in this setting the shifting shape of time is depicted as translucent clouds in an expansive animated atmosphere and a vague sheen of humidity dances through the air.

now the figure begins to flicker as if it were somehow composed of a black geometry of flame and smoke such that the trace edges of its profile twitch and vibrate wherever its form and weight meet space.

perhaps the figure represents a moving marker of generative motion—some self-defining process of an emergent ontogenesis …

[second scene for a theatre of one]

return to the previous scene.

the present object of intention is a figure whose contours periodically flutter, quiver, and shake, so that the figure centered in the reader's field of view projects a dark outline with an agitated profile.

the present object of intention has arrived in a foregrounded position, down screen from horizon and backdrop.

the atmospheric emptiness of the limitless expanse described and represented here as empty volumes and recesses now undergoes transformative change and the view from everywhere closes in to become another kind of frame.

now the minimal landscape morphs and becomes a black and white drawing of a blackboard positioned center stage in the front of a room.

concurrently, the figure with an agitated profile morphs to become an apparition of an animated tongue diagrammed on the surface of the black- board's blank space.

the tongue floats freely there. it wavers in empty space. and it is the height and width of speech. and it is the size and shape of volumes.

the tongue wiggles and moves itself over and around the flat blackness of space without meaning.

now, ever so slowly, sound begins to fill the room. the reader hears the whispers of a thousand unseen voices, the clatter and murmur of multitudes, speaking just to be.

[third scene for a theatre of one]

return again to the previous scene.

the present object of intention is an animated tongue.

the reader is staring at volumes of space rendered as flat depth in a dark room where a chalk white tongue floats against the surface of a blackboard's blank space.

and there are auditory sensations. there is mumbling. there are whispers and grumbling.

here and now, the figure of a chalk white tongue returns to a state of expressive agitation.

it begins to shiver. it trembles and flickers. it shakes and rocks back and forth violently.

now, as before, the present object of intention morphs and becomes—something other than itself. the lines and contours change and become the squiggles and lines of a symbol or a glyph.

perhaps the symbol represents the sum of both more and none, some intended sort of gesture, a sort of lexical reference to something unseen but nevertheless present.

[FIRST PART:

ON THE EMERGENCE

OF MEANING

FROM MATTER]

[psalm of the not-for-us]

sit down. place your hands across your eyes or tie
a blindfold around your head and place your hands
inside your lap. now lean back and breathe deep.
go back to the back of it. to the deepness inside
of it. to the not-even-black of absolute emptiness,
some faceless mask of nothing. no eye, no mind,
no spirit, no matter, zero: or rather non-zero in the
transfinite void of voids. and now a tiny light in
space. some tear or rip inside the formless void so
that without reason or purpose, the weight of the
world given in a sphere. and so time as paroxysm
from infinite density. red lights, blue lights, green
lights, and deep plasma pearl white speckled dot
pinholes bursting forth from singularity. the void
itself is dressed in the blue hue of absolute emp-
tiness, crooked electric contingency dancing in
an unimagined situation of perpetual deathsong
and rupture such that each successive combustive
disjuncture yields sufficient space for a process of
ongoing recurrent contingent emergence. and so
the birthing of both brightest storms and terror

in darkness, an ontology of erasure unspooling unscreened unmade film in lengths of measureless distance as serial unsung songs proposition the disrobed cosmos. imagine the in-itself of electricity and primeval fire. the in-itself of cold hatred of supereons and perpetual sleep waiting behind untraveled hallways of black confetti and dark matter. stars bobbing in the nothingness. death masks sleeping amid ionized air, resting amid the unmapped calculus of unwritten functions in disjointed distances for no one and nothing. there is no for-us. there is no for-us in broad arcs of time without beginning or end, darkness moving against the deep: the Crab Nebula, the Horsehead Nebula, astrophysical presences in the arc of night, spinning, expanding, collapsing, and coming undone. now the undisturbed sleep of the ancestral realm. the relics of the *arche-fossil* remain untouched. they are alone. they wait amid the silent expansive manifolds of our continuously unremarked, sentence-free absences.

[something from nothing]

maybe stasis. then motion. then motion in distance. or maybe movement from absence. black shivering kinetics of waveform spreading out as a humming-tinging-pinging vibration of sound. the sound shakes itself loose and wanders silent voiceless distances where energy is readied and bent into form. and now light floods into space as form extends itself, outward and forward, repeating and expanding as countless unnumbered extensions. and so the freedom of movement through empty time. and now serial motion down hallways of unexamined shadow. unmeasured space becomes a celestial cathedral for unspoken questions and rootless interrogations waiting to become riddles, and aporias, and mysteries, queries that hang in the air as shape and form grow outward while intensity reaches forth from the inside out. and now time yields a sphere, a ball, a proto-center, around which swirl, rocks, stones, and pebbles, a growing mass of dirt, gas, and sediment. it spins in the infinite grandeur of space without feeling. and

now a congealed core of iron nickel alloy. and also a
mantle of iron and magnesium rich silicate rock. this
growing extension gives itself form in jutting slabs
of massing granite. as uneven crust. as a *bric-a-brac*
of stone and rock rolled into a topological ball
situated and thrown into space for the unfolding
yet unobserved theatre of being and becoming.
and now rocks and stones clothe the spherical
planed space of *a thousand plateaus*. and there are
trees, rocks, and crags here. there is cosmic breath
pulsing in and against both distance and darkness.
fields and vectors of energy move of their own
accord. they vibrate beneath the gloaming teeming
air of periwinkle stars. and now the gold and green
divinity of rain. and now all the shine and weight
of divine minerals. metal, heat, carbon, and water
ready themselves and give way to breath without
thought. it exhales and circulates in boundless
magnitudes and broad arcs of geologic time where
stone monuments stand forth in primitive hollow
space. all hallowed hollow darkness. and now the
cruel quiet of birth. and now, the pure joy of death
and emptiness.

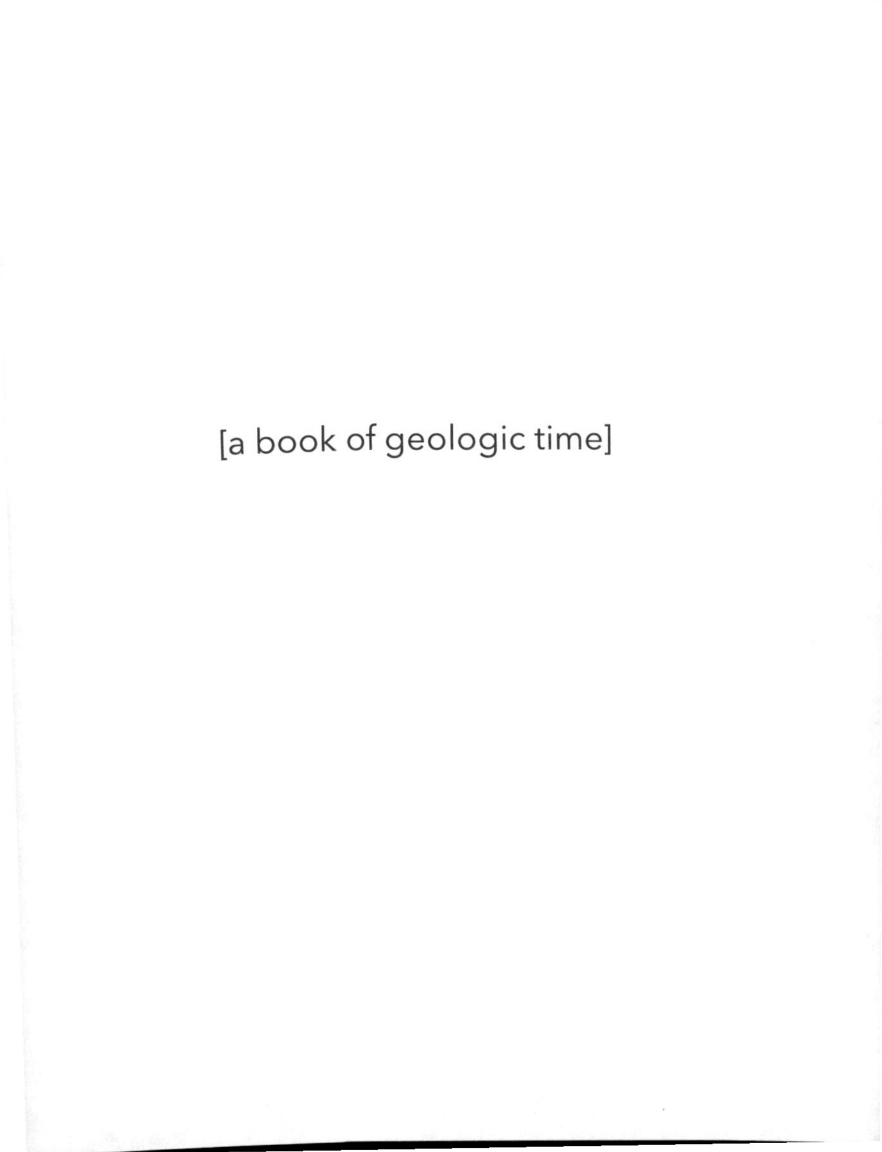

[a book of geologic time]

Costumed readers on lit-up platforms on two sides of a stage. The platforms are set back about halfway from the front of the stage. A large screen is set even further back and centered behind and above the platforms.

The stage space and the screen are intended for the deployment of visual and cinematic elements supporting the oral performance of the text. What types of visual elements to use and when to use them are left to the director and performers but care should be taken to keep the performance simple so as to reinforce the content of the text.

The left platform is for one solo reader. The reader on the left is dressed in robes and stands behind a pulpit. This is the alpha narrator.

The right platform is for the remaining (five to nine) readers who serve as beta narrators (analogous to the choir or chorus (the khôra). These readers are dressed in costumes taken from the text.

The text is divided into numbered sections further subdivided into sentences and paragraphs.

The alpha narrator reads the first and last portions of the first section, the first half of the third section, the eighth section, and finally, the last portion of the ninth section. All other portions of all sections of the text are read by the members of the khôra, who alternate their oral performances with each hard return.

Stage lighting follows the text from performer to performer.

begin again. before the before. amid inertia and drift. in high most reach. in the grip of depth, a circular, even spherical expansion of mass and energy as moment and instant become origin and immanence.

and now a view of Laniakea, the Virgo Supercluster, the Local Group, Andromeda, Triangulum, and the milky alabaster and peach white spiral arms of the Milky Way.

amid light and darkness, a small sphere comes into focus, illuminating the curves and contours of a molten, red-orange, tiger-eyed, black calico planet, superheated, and spinning on the cold glass floor of this, our galaxy.

now the age of Hades, the Hadean Eon, zircon crystals, rock vapor breath, and atmosphere congeal midair in the quiet void of the not-for-us.

here now the first ocean and possibly life but not

likely any life. and so another Eon, the Archean, follows. and so heights and depths and volumes and distances. amid these distances, sanukitoids and granite, mudstone, and other various protoliths, thrusting their form and weight upward and outward to bear forth, the totalized immensity of cratons in the coming-to-be in the age of beginnings.

and so igneous rock and deep water. photosynthesis, enzymes, and amino acids. electricity amid the gaseous atmosphere yields both organic soup and microorganisms as evidenced by stromatolites.

so now the Proterozoic Eon: the age of "earlier life." accumulation of oxygen in the atmosphere, increased subduction, marine transgressions, epicontinental seas and the supercontinents Columbia and Rodinia.

advanced single-celled life, Eukaryotes and multicellular life and the symbiosis of mitochondria and chloroplasts.

so-called and hypothesized Snowball Earth comes and goes as the Phanerozoic Eon opens with the Paleozoic Era and the Cambrian Explosion.

II

approximately 541 million years ago a sumptuous
explosion of carnal form, foam, and unseen color.
energy and wind and soil and water.

algae, invertebrates, shelled organisms, armored
arthropods, and trilobites swim in vast waters
where marine phyla evolve and give way to the
Ordovician Age.

and so forty million years of primitive fish, cephalo-
pods, coral, snails, and shellfish.

the Earth's crust slips and slides over and across its
own rock mantle, rearranging the continents as if
they were unmeasurable malleable puzzle pieces
evolving in shape even as life itself evolves in count-
less forms.

slowly the continent Gondwana (an amalgam of
Africa, South America, Australia, Antarctica, and
India) moves from the equatorial region to the

Southern Pole even as Laurentia (parts of Europe and all of North America) collides with the remaining parts of Europe known now, retrospectively, as "Baltica."

the first life on land begins and ends as a first mass extinction occurs due to the Earth's cooling which is thereafter followed by a period of warming and the subsequent advent of the Silurian and Devonian Ages.

and so the further evolution of fish, including the first freshwater fish. there is a radical diversification and hyper-multiplication of marine species even as processes of animal and plant diversification play out across the land: earthbound Progymnosperms, *Elkinsia polymorpha* and *Tiktaalik roseae* herald the birth of trees, the birth of seeds, and the emergence of amphibians and tetrapods. at the same time, in the space of Earth's oceanic waters, spiriferida, gastropods, ammonites, ostracods, sea scorpions, placoderms, and coelacanths.

for 60 million years the Devonian push and tug of circulating currents, the roll of foam and spume-bearing waves across pelagic depths and

over inshore shallows until the Age of Fish comes
to a close with a second mass extinction caused by
who knows what.

III

from whence comes life and where does it pass?
some say life is a great chain of being, that every-
thing is connected in a great hierarchy assigning
form and place to everything in time and space.

inert material matter, including rocks and dirt, form
the base of the hierarchy, while spirit beings, and
in particular, the one true God, occupy the apex of
the hierarchy.

the great chain of being is perhaps encased inside
an eternal four dimensional block of time, though
God, being subject to no one and nothing, stands
outside this block of time, observing the beauty
of his creation, the sculpted and totalized express
manifestation of divine purpose and will.

in other cosmological viewpoints life is understood
as a woven and interconnected energy field while
time is thought of as an expanding circle of intensity
twisted in flux.

recurring cycles of life and death manifest the *samsara* of the world's throbbing pulse actualized through kinetic processes of biochemical emergence, extension, and entropy.

life begets life begetting death and more death begetting even more life in an ongoing never complete cycle of immanent emergence.

and so peaks and troughs of virtuous circles of life-giving activity counterposing themselves to and joining with peaks and troughs of collapse, death, quietude, and seemingly endless inactivity.

IV

in darkness *ouroboros*, the world serpent, swallows
its tail. and water flows, and spills, and splashes over
rock. and water splashes over stone and sand until
water eventually regains a greater quantity of life.

and so the Carboniferous dawns, the so-called
Age of Amphibians: massive arthropods, tropical
swamps, the appearance of amniotic eggs, reptiles,
synapsids, and the further spread of trees and ferns.

the Appalachian and Ural mountain systems pull
themselves from the ground, standing to new
heights as the Earth's shifting tectonic plates wres-
tle and quarrel with one another.

now, for a time, deltas and flood plains. warm
shallow marine ingressions, massive insects, rich
oxygen atmospheric conditions, and the laying
down of peat and coal deposits.

for six hundred thousand centuries Gaia blossoms

as moss, sweating the sticky fruit and green heat of
the dank fen.

now slippery, scaled, and more smooth-skinned
temnospondyli radiate in every direction, silent,
multiform, reticent, unobtrusive salamanders,
carrying their elongated bodies and webbed feet
into an untold, folded, multitudinous diversity of
terrestrial, aquatic, and subaquatic habitats.

in time the humid tropical climate of the Carbonif-
erous gives way to a more arid order, which, in turn,
leads to the glaciation of Gondwana.

and so the Carboniferous Rainforest Collapse, a
subsequent period of habitat fragmentation, insu-
lar biogeography, and pervasive endemism.

V

time moves. another break and another shift. the Permian commences with widespread glaciation and the ongoing consolidation of the Earth's major landmasses.

now the supercontinent Pangaea rests in the universal all-sea Panthalassa, a vast ocean harboring and reflecting the lights of the midnight cosmos over its waters so that there is a rippled shimmering reflection of eternity dancing and skittering across the waves of Panthalassa's unending depths.

and now a limitless continual desert spreading its orange, ochre, white, and crimson sand to the far reaches of Pangaea's outstretched sprawling distant surfaces.

and so a flourishing of better adapted conifers, reptiles, and the advent of ever-newer synapsid forms bearing the differentiated teeth, jaws, and temporal fenestra, portending the future rise of clade Mammalia.

after a 45 million year secular rise in temperature,
the Permian terminates, as a consequence of a fun-
gal spike, climactic change, volcanic overkill, and
oceanic acidification. now the Great Dying. seventy
percent of vertebrate terrestrial species and ninety
percent of marine species die off.

VI

language can be both precise and evocative. symbols and words can both denote and connote. take, for example, the manner in which symbols mediate how we think about, approach, and respond to the allure of a spiral.

the whorled chambers of a nautilus, the petals of Neptune roses, the watery vortices of cyclones and maelstroms arouse and beguile us, impressing on our minds the hypnotic imprint of nature's own peculiar typeface.

in the desire to penetrate the spiral's mystery, mathematicians delineate the essence of the spiral's form. a spiral is defined by the line that flows circularly outward from its center, the origin, the locus around which a spiral's arm wraps itself. these features can be expressed in formulae that designate the precise relations of a spiral's constituent parts:

$$r = a\theta \quad r = ae\theta \cot$$
$$\text{or alternatively,}$$
$$r(\varphi) = A/\log[B \tan(\varphi/2N)]$$

yet one may also use symbols to imbue a spiral with a more evocative yet less determinate significance.

as one of the earliest and most widely used sacred symbols, the spiral signifies the journey and emanating metamorphosis of nature through space-time.

time's cycles are the concentric circles bent around the spiral's originating center while the length of the spiral's arm (or arms) represents time's duration and the primordial power of nature to reach forth from the inside out.

picture the arm (or arms) of the spiral moving in tandem with the unfolding of evolutionary processes across the ages of the Earth.

picture the spiraled double helix of life's DNA vibrating as source and energy.

the Earth grows and expands. it replicates and transforms itself where form and weight meet space.

VII

now the aftermath of the Great Dying. the Pha-
nerozoic Eon continues in its ellipsis. the Paleozoic
transitions into the Mesozoic while in succession,
the Triassic, Jurassic, and Cretaceous periods open
and unfold.

and so 180 million years of dinosaurs.

the Triassic, sandwiched between two mass extinc-
tions, begets beetles, ichthyosaurs, grasshoppers,
stony corals, and the Triassic lizard kings *Isanosau-
rus*, *Prestosuchus*, *Alwakeria*, and *Coelophysis*.

now the deep complaint, the bellowing and rifting
of the supercontinent Pangaea. the land pulls itself
apart to form the continents Laurasia and Gond-
wana. Laurasia is an amalgam of North America
and Eurasia. Gondwana is still the puzzle of South
America, Africa, Australia, Antarctica, and the
Indian subcontinent fused by the machinations of
the Earth's folding, grumbling, mumble-some crust.

now pine forests appear. and the first ancient croc-
odilians. and the first mammals. still, this is princi-
pally a time of reptilian activity when birds evolve
from theropods to fly above the angiosperms of
the flowering Cretaceous which closes when an
asteroid slams into the Earth, yielding the great
Chicxulub Crater, the 5th mass extinction, and the
K-Pg boundary. now the death of dinosaurs and 75
percent of all life.

VIII

what is narrative? what is it for? where does it come from and what does it imply, if anything, about the nature of meaning?

what are stories? do they matter? if so, why and how do they matter? is a story a simple organized recounting of events? or do stories unmask latent purpose behind the events that constitute the elemental pieces of the story's unfolding plot lines?

perhaps narrative enables thought to grasp that which is unseen.

or perhaps narrative allows people to organize time and matter with the promise of reason, intention, and foresight. in this manner stories might-could-maybe fashion a world that can be conceptualized as an inten-tional whole.

or maybe not. maybe stories only offer empty prom-ises of underlying meanings that they cannot supply.

perhaps stories conjure not only that which is unseen, but also, that which does not exist.

IX

now the Cenozoic. the Age of Mammals. the Age of
New Life.

Earth's temperatures rise for ten million years, first
during an extended period of gradual warming,
but subsequently, more rapidly, as a consequence
of carbon releases triggered by volcanoes, seismic
shifts, and the freeing of quantities of methane from
tectonic sea beds.

picture the Earth ablaze with heat and light as tem-
peratures reach a geothermal maximum that peaks
at an average temperature 14 degrees Fahrenheit
greater than our present average temperature.

the heat is followed by planetary cooling, extended
periods of climactic fluctuation, another period of
warming, and a subsequent period of further cool-
ing, leading to the glaciation of Antartica, the spread
of ice sheets, and in time, the advent of the Pliocene
and Pleistocene.

now the continents begin to take their present shape. the North Atlantic opens and the Tethys Sea closes. India pushes into Asia and the Himalayas, Alps, and Rocky Mountains rise from the ground, standing amid the slow motion violence of stone against stone.

placental mammals appear. grass evolves. grass expands. and grass gives itself over to great grazing herds of hoofed herbivore fauna: eohippus, the dawn horse, and great antler-crowned deer. and also elk, and Gemsbok, alpacas, guanacos, and the Bles-bok, as well as strange forgotten camels, antelope, impalas, the greater kudu, the kongoni Hartebeest, wild sheep, and cows, and wildebeests, bison, goats, and gazelles.

new apex predators occupy ecological niches opened by the death of dinosaurs: the giant short faced bear; marsupial lions; *Smilodon*; 50 ton orca *Leviathans*; and Megalania; and Entelodon, the hell pig. great terror birds such as *Paracrax* wander the land feeding on less able life forms even as whales and elephants appear alongside dogs and cats.

and so the world becoming what it is…

the planet now witnesses the beginning slow emer-
gence of our forebears from some common chimp
and human ancestor.

[descending the tree of life]

once there was a monkey, yes. once there was a monkey-ape, yes yes. once there was an ape. a big hairy ape. one two three four. once there was an ape. yes yes. once there was an ape-man. a sort of proto-human ape-man. five six seven eight. it ate roots and fruit. it ate bugs and nuts. in the early morning, and at night, and all the live long day, the apeman was eating roots and bugs and nuts. and the ape ate leaves. yes yes. he ate roots and bugs and bugs and nuts (but also leaves and fruit). one two three four. once there was. once there was a dark shadow laden forest. and in the dark shadow laden forest there was a short and stocky woman-ape. she would climb and climb. yes yes. she would swing and climb and clean her teeth. yes yes. one two three four. one two three four. once there was a forest. five six. once there was a forest bounded by savannah. yes yes. and there was water. and hills and trees. overhead, the movement of the Sun continued. five six. seven eight. once there was a place—a place the woman-ape and apeman would gather when they roamed. and they would roam. and stop. and go again. and once there was a place. yes yes. and the, one two three four, woman-ape, and the, five six seven eight, apeman would gather. and they would join together with others of their kind. and they would join in small collectives. and they would sit. and they would sit and stand. and they would walk and they would climb down and up. and they would hang from trees. and they would swing through the branches of the trees. and they would climb. and they would *walk*. once there was a place they would sit and stand and climb and once there was a place they would gather. and the woman-apes and the apemen, they would walk.

[a field guide to bipeds]

The texts that follow provide the written content of thirteen placards placed in the wing of a museum that features thirteen dioramas.

The dioramas contain a three dimensional sculpted depiction of a single hominid species amid its habitat. The placards are attached to the walls adjacent to each diorama.

Alternatively, the texts are written on index cards that accompany a set of goggles and a pair of ear buds. An individual can view and experience the quotidian world of each hominid species featured in the texts using the goggles and ear buds.

"Toumai"
(*Sahelanthropus tchandensis*, 7–6 MYA)

Black star of dawn. *Hope of life.*

Invisible below the skull, the centrally-placed foramen magnum, face, and cranial brain case float above phantom bipedal legs.

Liminal. A border crosser. A crypto-hominid moving through woodlands and shadow.

He peers and he stares. He gazes at nothing.

"Millennium Man"
(*Orrorin tugensis, 6.2–5.8 MYA*)

Fossils BAR 102'00 and BAR 103'00, both femora, indicate that

Orrorin walks upright, proud, scattered, searching. Yet with

long arms and curved grasping fingers he still moves about the forest, rummaging around—for the symptoms of his life.

Ardipithecus kadabba
5.8–5.2 MYA

Pieces of jaw, hand, arm, and collarbone. A few teeth. Another tooth. Other teeth. The species-defining toe fragment. Narrative infers so much—from so little.

"Ardi"
Ardipithecus ramidus
4.4 MYA

Ardi's divergent opposable big toes make her feet appear as if they were another set of hands.

She ascends through the canopy. She climbs up. She climbs down. She climbs up. And down.

On two legs, she moves across the ground, measuring with each stride, all the places she will go.

"Lucy"
Australopithecus afarensis
3.9–2.9 MYA

About 70 footprints like ellipses through volcanic ash. The pelvis, knees, legs, spine, and feet signify a swinging gait, upright posture, and the onward forward roving arrow of travel.

Australopithecus africanus
3.3–2.1 MYA
(Taung Child)

The hand bones suggest the ability for precision-gripping tool use though dig sites associated with this species remain bereft of tools.

A. africanus stands on the stage of time. She tucks her hands behind her back and walks away backwards, hiding what she holds.

Homo habilis
2.4–1.4 MYA
"Handy Man"

Rock choppers, hammer-stones, and lithic flakes.

Homo habilis pounds and shapes stone implements in a creek bed or incidental rock quarry, the clack clack of quartz or basalt flooding the air with a new human heartbeat.

Homo ergaster
1.9–1.3 MYA

The Acheulean bi-faced stone handaxe. The larger braincase, shorter arms, and longer legs. In search of greater bounty.

Homo erectus
1.9 MYA–50 KYA (or)
0.8 MYA–50 KYA
"Upright Man"

Probably the first human to range beyond Africa, possibly as early as 1.75 million years ago. Mixed evidence of fire use at 1.8 MYA and 500 KYA, though clear evidence of intended fire use around 400 thousand years ago.

H. erectus traveled far and wide for nearly 20,000 centuries.

He was both a primitive Odysseus and the original Prometheus, but who did he worship? How did he pray? To what gods did *H. erectus* owe his superior longevity and when and where did he steal fire from heaven?

Homo heidelbergensis
700 KYA–300 KYA

The first human to build shelters. Perhaps the first to bury its dead.

Homo neanderthalensis
400 KYA–40 KYA
"Neanderthal"

Neanderthals used flint, tended hearths, painted cave art, made ornaments, and fashioned bone tools before disappearing from the fossil record 40,000 years ago. Where did they go with their oversized faces? How did it end? Did they close their big round eyes forever, after witnessing the first human genocide?

Homo denisova
400 KYA–30 KYA?

A jawbone found on the roof of the world.

A pinky finger and three molars from the Denisova Cave in the Altai.

Like other humans, the Denisovans lived an endless nomadic existence. Until they didn't.

Homo sapiens
200 KYA to present

Use language. Make art. Believe in gods. Believe in spirits.

Responsible for nuclear fission reactors and silicon microprocessors.

Humans can clone animals.

People can clone people.

We set the world on fire.

We put a frame around time.

[being and time]

I

the seasons turn. the days pass into night even as night passes into day.

and the years come and go and come around again and there are sky-arcs through emptiness and there are pathways and iterations of travel.

there are patterns of light and patterns of shadow, mimicking form.

there is the rise and fall of heat and the advance and retreat of cold.

in the evening, at night, and in dawn light, the Moon puts on and erases its face.

and so life multiplies. and life grows. and life flowers becoming ever more multivalent, blooming, and multiform, so that in time, there are dozens, then hundreds and thousands of eyes, taking it in.

and so groups and bands of nomads with their

pupils peering skyward, the eyes of each and all of these observing the wonder of the great celestial mechanism, the cosmic motor works, learned and mapped variations in the length and width of days, the decorated habits of Earth, sky, and solar system.

and so a field of eyes open, staring, watching from a locus of finitude.

the eyes move about the free space of absence amid the emptiness of the eye socket.

they turn and look and wheel around.

they turn and look and look again, observing the gyre, the awe, the marveled bewilderment of just being-there, without reason or purpose.

II

picture a clearing in a forest.

picture a woman seated in the clearing.

now imagine an invisible box of habitable space placed and stretched all around her like a rectangular prism forming an invisible whole body halo.

now move inside this unseen orthotope.

seat yourself inside her seat inside this box and peer out into the framed three dimensional picture that projects outward from the depths inside her head.

you can almost feel or see how she somehow carries with her, a mobile stage of play.

here then: a picture of experience, a box of time.

in this box of time, a composite image of the velocities and inertias and trajectories of known sequences of human deliberative motion intersecting with the known velocities and inertias and trajectories of the careening habitable planet.

how to describe it? what is seen and felt. what is known and experienced. what is learned, remembered, and handed down across time.

the movement of great and tiny clouds pulled across variable topographies of dirt, rock, and sediment.

and the circulation of the four winds.

and the dripping precipitation of periodic events of precipitation.

as such, the drift and inertia of sprawling atmospheric knotted outstretched weather systems.

and so, episodic or predictable events of rain and snow and sleet and hail.

and the falling of leaves and the rising of humidity.

and also the gathering of great herds of reindeer. and aurochs. and bison and elk. and all the polychrome wild horses.

and the serial gatherings of ibexes. and hyenas. and mammoths.

and great grey, white, or black wooly rhinos.

and billowing great fires.

and jagged white-blue streaks of lightning, knifing through the air to cut open the panorama of it all, spread out, right there amid the complete embrace of so much felt distance.

into and through it all, the nomad peoples of the Earth migrating and moving themselves over and across land bridges of unwritten history.

the daughters of Mitochondrial Eve move out of Africa and across the Levant.

they move over mountains, across rivers, and hills, and into the wide open long flat and rolling plains of Asia.

they move across great bodies of water and into the islands of Oceania.

they move through the valleys and forests and hills

and the ice fields of Europe.

and still the spinning Earth.

and the movement of the stars.

and the continual overhead dance of the planets adjacent to the always changing shapeless face of the equivocating Moon.

through it all the presence of each human wheel-eye eye-wheel frames the presence of that which is absent yet present in memory, organizing for the mind, the ground and horizon on which the human nomad carries itself, traveling, in never-ending pilgrimage.

III

how long ago it was. how very distant and hidden yet reachable.

the evolved cognitive apparatus of our advanced primate minds allows us to reach back through unseen time and grasp the meaning of certain milestones.

four to five hundred thousand years ago: a beach in Asia, an etched shell in Java, and on the surface of this shell, a zigzag crosshatch geometric representation of maybe waves.

perhaps these waves signify the right angles of time, the beat and exacting angular pulse of the variable instant, its comings and its goings, scratched out in broken linear forms carved into, over, and against, the scratched and altered surface of calcified *Pseudodon*.

IV

one, two, three, or four hundred millennia pass.

one or two or three or four thousand centuries pass.
that's between one hundred and four hundred
thousand years.

beads appear, red ochre pigments, and objects
scored with dots and lines and hashmarks.

framed in the walls of earth-sky-fire-time, fabricated
meaning reaches for the unseen.

V

round stone or bone or hard-worn ivory and serpentine.

carved black jet or worked antler or fashioned hematite.

molded ceramic and/or the molded limestone likeness of the female form rendered rounder in the depths inside the darkness:

the Venus of Hohle Fels.

the Venus of Lespugue.

the Venus of Willendorf.

the Venus of Moravany.

the Venus of Monruz.

all these faceless figurines rendering a red image of the self bathed in the blood of time.

VI

blinding light. and darkness. and hunger. and the urge of lust. and episodic and routine and even uneven searing incidents of pain. and now and again, moments of physical release. and also moments of mental release. these perhaps followed by periods of openness and clarity. and also joy and sorrow. and the always overhead comings and goings of the celestial clockwork set against the spread out too distant heavens. and so the Sun and Moon. and clouds and stars. and the iterated repeat changing shape of nighttime shadows, each and all of these splayed across the surfaces of earthly human habitats. but also now the continual movement of limbs into and across the day. these limbs making their way over and across the land and all its wanton wild bounty. and so the continually iterated reach of arms and hands and fingers. toward objects of necessity. but also towards objects of desire. concurrently the opening and closing of lips and mouths. and the gleam and flash of teeth (in the Sun) but also the gleam and flash of teeth before flames and nightly studied fire. somewhere along the way early humans come to occupy mountainous or stony hillside hollows. they climb inside a hole inside a hill or mountain. and the mountain grows or the mountain shrinks. or the mountain grows and then shrinks and expands or the mountain stays exactly the same size for just one second or even maybe four or 40 centuries. and then, little by little, serial observations followed by the intentional representation of known and habitable environs on the walls and ceilings of earth-stone dwellings. layers of images now appear and multiply on the cavernous surfaces of here, and there, and there. and now grey and red and black images drawn against hard cinematic stone. there are white spirit images drawn on the surfaces of flat or rounded space and ochre diagrams on top of ochre drawings on top of ochre paintings yielding the side by side representation of known objects with a picture of something somewhat different, some sort of represented ethereal felt but unseen "otherness," a thing imagined and invisible, a kind of strange present emptiness lurking below the surface of that which has been seen and experienced, so that, here and now, the emergence of a composite-real yet also spectral-unreal manifest image of time's present place of being. multiple layers of pigment become multiple layers of inferred meaning. and suddenly there is an untold bounty of red hands appearing on the surface of a wall in Sulawesi. and half a world away there is a multitude of red hands on the painted walls of Chauvet. and there are white-grey rhinos massing inside a motionless stampede of time. and the brown bison of Altamira and the stags and aurochs of Lascaux now become, over time, the heralds of some dancing horned sorcerer rendered real and made manifest in acts of ritual commemoration rooted to the rhythm of the Earth.

wheat from the ground. and wheat ground into flour. and human migration slowing so that there is a settling down of many (but not of all). as such grain is planted. and grain is gathered. and grain is stored. and so transactions and accountants and record keeping. but also the rules of priests and gods. and the learned explication of the divine order of things so that there is an elaboration of purpose. and a securing of position. from top to bottom there is an ordering of things. and so hierarchy. and structure. and the constructed articulated elaboration of duties as well as the allocation of place and space and income. as such theocracy and pottery and megaliths. but also marks on stone and clay and bark. marked lines on wood. and linen. strange shapes on palm leaf in the palm of the hand or marks on parchment placing the found power of script against leather, silk, and vellum. prehistory becomes history. and *everything is alive*. the ghost and the shekel. grass and rocks and trees and clouds as well as the will of the ancestors. and the will of gods. and the will of angels. and devils. and other divine beings. wolf spirits and bird spirits and snake spirits. in a certain sense written language manifests the will of these spirits even as the power to write grows up alongside both accounting and the practice of divination. lines and squiggles and circles come to compose the symbols, sigils, and glyphs written (like animal tracks) across the void of time. and so pictographs, ideograms and hieroglyphs, giving way over time to graphemes and alphabets: Jiahu signs on tortoise shells; Banpo symbols; the Dispilio Tablet. also Vinča symbols on pottery. and Sumerian pictographs. and proto-Elamite. and Egyptian picture-words for dog-gods. picture-words for cat-gods. and Sun-gods. and Sun-god words. and Indus Valley Script. and Akkadian cuneiform pressed ever so firmly into soft clay tablets, the tablets made and fired in kilns by temple scribes taking down the divine sacred words of priests or priestesses. but also now Elamite cuneiform. and Linear Elamite. and Cretan hieroglyphs. and the proto-Sinaitic alphabet. and Hittite cuneiform. and Linear A and Linear B. and the Ugaritic alphabet. the Phoenician alphabet. Chinese oracle bone script. and Paleo-Hebrew.

and Archaic Chinese. and Olmec and Greek and Aramaic. and Etruscan and Latin and Zapotec. and Mayan glyphs of organized astronomer priests. and Nabataean. and Coptic. and Runic. and also Arabic and Hiragana. and Glagolitic. and Katakana. and Cyrillic. written language. and the rendering of it. initially, by those with specialized knowledge concerning both spirits and money.

[the sum of it:
first reprise]

energy. and earth. and breath. and thought. these composed and placed atop the surface of the page not for a sense of juxtaposition, but rather, with a relationship of extension in mind.

think of stones arranging people in time. think of the Earth thinking you into its sequence.

there are fire and honey narratives. there are cloud narratives. there are also wood and water narratives. and therein storms will drift. or float. ideas will float in ether above or amid stories of folded paper airplanes.

and so ideas and images, combining to fit the page like a frame drawn around the empty nothingness of now, the shape of the frame, somehow expressing and encompassing life, even as life just goes right on happening.

[Schelling, Spinoza, Bergson]

maybe nature creates its soul as fiction. or maybe matter composes itself in thought. I'm not speaking about the dead weight of the actual. I'm speaking to the real pulse of the possible. the possible meanings transacted and cut up inside of you. inside of your secret. because believing is more vital. because believing is more or less sensual. it feels like a hundred doors being opened. it feels like something you can feel but don't see.

[SECOND PART:

ON THE PERFORMANCE

OF MEANING

BY PEOPLE]

[Prospero
speaking amid the
residue of time facilitates
a narrative bridge
to the present]

A dim theatre. Center stage, a large oversized table covered with scrolls and books. Upstage, behind the table (center stage left), a large chalkboard facing the audience. Further upstage, in the background, positioned from left to right, the looming silhouettes of great and famous monumental constructs: the Pyramids, the Taj Mahal, the Parthenon, the Coliseum, the Great Wall, the Eiffel Tower. In the space between the architectural wonders and the area containing the table and chalkboard, massive bookshelves filled with books. In front of the bookshelves, center stage right, a tall spiral staircase descending from the heights to an area just behind and adjacent to the table.

In the background a flash of stage lighting evokes the likeness of lightning, and for a few brief moments, the audience hears wind and thunder. Now the magician Prospero, dressed in robes, descends the staircase to seat himself at the table facing the audience. He looks through the scrolls and also consults the books strewn about the table. From time to time he pauses to scribble something down in a folio opened in front of him. The audience hears him mumble.

After a few minutes Prospero rises from his seat and begins to pace. He approaches the book shelves, periodically removing and consulting books, though he almost always replaces them after reading a passage or two. Sometimes he places a book from the shelves on the table. Following a few minutes of this he perks up and walks downstage, approaching the audience to whom he speaks directly:

Language and spectacle. Performance and Illusion. The magic of it. The magic of rendering the unreal real with the learned artifice of words…

Prospero, using sleight of hand, seemingly pulls a bouquet of flowers from the air.

I, Prospero, deign to instruct thee [*while still holding the flowers Prospero points to the audience*] in the unlearned secrets of what is…and what is not.

Prospero makes the bouquet of flowers disappear.

With language, the power to point and gesture *[Prospero points and gestures]* becomes the power to render myth with a brush, or a stick, or a feather, or a quill. *[Prospero pauses briefly and then walks to the chalkboard where he begins scribbling sigils and glyphs.]*

Let us speak of the development of mythology, religion, and the Law…

The Law is recorded and set down to represent the very frame of everything, the divine will of the Law-giver. Storytellers make it so with epic explanations, framing the values of an age. And so:

> Gilgamesh.
> and Hellan.
> and Moses.
> and the Mahabharata.
> and Romulus and Remus.
> and the Shahnameh.
> and Beowulf.

and Scheherazade.

and Siegfried.
and Roland.
and Don Quixote.

Smoke or vapor begins to rise in the backdrop, slowly and progressively creeping downstage as Prospero addresses the audience again.

Yet all of this as *da Sein*. As *there-being*. As human being.

The human animal moves over and across the Earth *taking-the-this* by means of the elaboration of lexical reference and the growing capacity for linguistic recursion.

Prospero now returns to the chalkboard and begins writing again.

Hear then…the contemplation of the Cogito and…a new found concern with intentionality…and the expression of the so-called…"free will"…

Yet, we must also speak here of the presence of the unconscious by which I mean…*the Law of the hidden presence of..."the Other" as well as...the inverted ladder of the law of desire.*

Prospero erases everything written on the chalkboard and turns to face the audience.

The human being seeks integration with *the all*. Seeks self placement amid purpose and meaning. In this way the human's desire becomes manifest in and constituted through speech and the ability to write. To write it all down. To make sense of it.

Prospero turns his attention back to the recently erased chalkboard and begins writing again.

And so doctrines and...doxa...and...schema...but also...the emergence of *power/knowledge* and... *social hegemony* and...*bio-politics*...That is to say... without always consciously knowing it, we come to encounter ourselves as...*interpellated subjects* of *the Spectacle* of the *hyperreal*. As such we mis-recognize ourselves even as we move our dis-adaptive selves through the day, our bodies unwitting bearers of structure, oblivious participants to our own immersive subjugation to the dictates of *the law of the Father...the symbolic order of an alienated and machinic big Other.*

Prospero turns once again to the audience and gestures expressively before performing one last magic trick. He then addresses the audience one last time.

Who made this happen? How did it happen? Why did it happen?

The theatre set slowly fades from view while smoke and vapor roll downstage and out toward the audience.

[language as a vocation]
(intended for film)

The monologues that follow are performed back to back, on film, as a series, by the same actor in the same room.

Initially the setting for the monologues is a well lit white room with a white door in the back wall. The door opens onto the room stage right.

The featured room is completely empty. All the walls are white. The ceiling and floor are white. A long white string hangs from the ceiling in the center of the room.

When the action begins the door opens and a player dressed in black and wearing a Kyōgen Saru mask enters and looks around.

The player exits through the door briefly and then immediately returns, pushing what appears to be a pedestal (or soapbox) which is moved to the center of the room.

The player climbs atop the pedestal. The player stands on tip toes and reaches up to pull the string suspended from the ceiling.

The room goes dark.

The player pulls on the string again.

The room partially lights up to show an interrogation style lamp suspended about ten feet above the pedestal.

A cone of light from the lamp illuminates the space around the pedestal (and the player on the pedestal). The rest of the room is shrouded in darkness.

The player pulls on the string again. The room goes dark.

From this point forward the actor will perform the monologues that follow, bookending each performance by turning on and off the light inside the room. In most instances this occurs in the manner just described. The player stands on tiptoes and reaches up to pull the string. The room goes dark.

After a few moments the player (in darkness) reaches up and pulls the string again. The room then partially illuminates as a cone of light washes over both player and pedestal.

With each illumination and subsequent diminution of the light inside the room the audience sees that the player's costume has changed. Costumes, props, and performance notes for each of the monologues are as follows:

The Priest, dressed in robes, wears a miter, and holds a crosier. The Priest gestures the way a priest gestures—arms are raised and lowered in a manner that suggests enactment of ritual.

The Poet, dressed as Anne Carson, wears Anne Carson's Oscar Wilde suit.

The Philosopher, dressed as Socrates, wears a toga.

The Lawgiver, dressed as an Italian Renaissance prince, holds a scepter in one hand and a dagger in the other.

The Librarian wears a corduroy suit, sweater vest, and eyeglasses. A bookshelf stands behind the

*pedestal from which the Librarian recites her mono-
logue. When the monologue is complete and the
lights are extinguished, the bookshelf is wheeled
away.*

*The text of the monologue "the Censor" is not per-
formed from the point of view of a censor, but rather,
from the point of view of one resisting censorship.
The player is completely naked. Alternatively the
player is naked but wears a Pussy Riot mask pulled
over her head. The actor effectuates body language
suggesting interaction with a dominating authority
figure, alternatively resisting, submitting, and avoid-
ing the authority. As such, raised hands and cower-
ing activity juxtaposed to expressions of defiance
(such as a raised fist).*

*The Bureaucrat is dressed in a grey suit with a grey
shirt and a grey tie and is seated in a chair behind a
desk. In this scene the Bureaucrat neither turns on
nor extinguishes the light. (The light is turned on
and off by stage hands following a predetermined
protocol).*

*The Prophet wears a loincloth, has a long beard,
and is dirty. The Prophet prophecies from atop the*

pedestal, affecting a sense of agitation, both fidget-ing and shaking while speaking.

[the Priest]

the truth of which I speak is mostly invisible. if you close your eyes and pray you might see what you must surely feel, but only if you have faith, and only if your faith is sufficient to render you worthy. those who are called can see the blessed truth here illuminated before you. if you see what I say, raise your hands. now give thanks and praise and go out into the world to spread the word. access to the unseen is your reward. *many are called but few are chosen.*

[the Poet]

Homer and Hellan. Sappho and her lovers, flick-
ering. flickering Ovid and his lovers, frolicking. or
Eshelman before a campfire before the entrance to
a cave in southern France, pondering the

aurochs of Lascaux. or Jeffers astride Pacific cliffs
carrying Tor House stones while peering through
double cocked eyebrows at the void of the Pacific.
Inger Christensen stands beneath an

expanse of black sky watching the lit heavens blink
an eternal code of cosmic typeface. and Rexroth.
Rexroth with his thumb out, moving around post-
war Europe. or maybe Rexroth holding

his furrowed brow in his hands while pondering
truth in an ideogram. imagine Pound standing in a
cage, muttering while committing the words of the
Pisan Cantos to memory. and so

rain and Noh drums. and now blood pounding

through Pound's temples. Maximus Olson, a giant, wrestles with Melville the way one tectonic plate wrestles with another. and Satyrs and Whitman

making love on leaves of grass. or Vasko Popa and Ewa Lipska, mining a soul eternally subject to surveillance. Leslie Scalapino takes a picture of the present inside a frame narrative

inside a second frame narrative depicting a group of nomads cutting up all the mapped and unmapped territory of the eternal constant present. what is language in the hands of a poet?

ask Plato. or Badiou. ask Aquinas. or Heidegger. ask Agamben. or Kristeva. or Cixous. ask anyone. or everyone. ask yourself as you move around the planet, a newspaper in one hand, your death inside the other.

[the Philosopher]

why am I here?
what came before and where
did it come from?
what is the nature of being?
how do we know what we know?
how do we know what is real?
is the universe one or many?
do I have choice or is this an illusion?
where do I go when I die?
what do I owe my neighbor?
who should rule?
what is just?
should I kill myself?
what is the point?
what is love?
is love worth loss?
what is art?
what is language?
what is god?
does god exist?

is this a dream?
what is consciousness?
will I love again?
did my lover ever really love me?
will my lover return?
why am I here?
what is good?
what is happiness?
what is truth?
or beauty?
how should I live?
is it better to love and lose or
to never love at all?
do I have a soul?
what is thinking?
what are numbers?
what is time?
what should I do with
the time now given to me?

[the Lawgiver]

I am the Prince. a black finger of stone rising out of the earth bears witness to my great and glorious deeds. so say the gods. and I am here to defend the weak from the strong. and I am here to enrich my people. just look at how you have profited under my direction. and recall how I have protected you from the wicked and united you in your cities. recall how I have built up the temples. and slain thine enemies. and protected the church. thou shalt. thou shall not. for I am the law. I am the law. I am the law. *we find these truths to be self evident. for it is written.* and except as otherwise provided, *man is wolf to man.* but I am benevolent and chosen. as is evidenced by the wealth bestowed on me. so notwithstanding any other law or custom, the rules are the rules. for whomsoever: *ceteris paribus.* because we are a nation of laws. and no one is above the law except the lawgiver, who is the guardian of the state. he is justified in all acts. this, *the state of exception,* is willed by the spirits of our ancestors. and the divine right. for it is written. because because because. *for the health of our soul and our ancestors, and heirs. pray for us now and in the darkest hour.* though the people be free they shall also submit. *otherwise flogging, caning, imprisonment.* so say the twelve tables. for each shall bear their own sin. for whatsoever it profits a person, that person shall bear witness to the word. which came down through time. from the elders. and the priests. and all the great families. who are chosen. so says custom. as such I grant you the mercy of my eternal selfless love for you. *L'État, c'est moi.*

[the Librarian]

hello and welcome. I'm here to help. I believe the short text you seek will be found, pressed between the pages of a book, slipped inside a shelf resting against a large archaic brick wall, in the back of an even larger reading room, set off to the side of the main gallery. the text will be written in the future tense and it will refer to long rows of shelves containing stacks and stacks of books. it will describe the stone or wood architecture and open spaces of the library, as well as lesser known nooks and crannies where lovers pass notes even as they become entangled in the musings of a writer with whom they have only recently become familiar. the text will allude to broad expanses of time and the coming to be of words, and by extension, the creation of worlds using the dark powers of language. and so the celebration of signs. and symbols. but also references to the grandest libraries of the remote past and distant future. here you will find references to the Library of Ashurbanipal. and the Library of Alexandria. and Pergamum. and the Villa of Papyri.

and the House of Wisdom. and the Imperial Library of Constantinople. and the Bodleian. and also the Bibiothèque Nationale. and Trinity College. and the Beinecke. and all the grand national libraries of modern, past, and future nations will be referenced in the text. you will encounter all of this on page 114 of a little known treatise on cosmology, language, and nihilism. the treatise takes as its object the essential becoming of spirits from nature. you will discover the unseen truth of all of this on the floor of reading room 17, your finger pressed to your lips as you pass your eyes over the naked imprint of the words you hold inside your hands.

[the Censor]

if I can finish it, this poem will be a poem. if I can
finish it, this poem will be a poem about silence. if I
can finish it this poem will be a poem about silence
and noise. if I can finish it this poem will be a poem
about silence and noise and misdirection. as such
it will be a poem about obfuscation and truth. and
truth claims. and truth under erasure. and so it will
be a poem about power. and who has the power.
to name and speak and say. as such it will also be
a poem about who does not have the power. to
name and speak and say. in other words, it will be
a poem about Texas and Florida School Boards. if
I can finish it, this poem will be a poem about Ron
DeSantis, Vladimir Putin and Pussy Riot. it will be a
poem about the Stasi. and Lorca. if I can finish it,
this poem will be a poem about Lorca and Franco.
if I can finish it, this poem will be a poem about
Pinochet and Victor Jara. if I can finish it, this poem
will be a poem about Victor Jara's hands. if I can
finish it, this poem will be a poem about samizdat. if
I can finish it, this poem will be a poem about book

~~burning and witch burning. and Franklin Graham.~~
~~and Jerry Falwell Jr. and Liberty University (which~~
~~is not free). if I can finish it, this poem will be a~~
~~poem about Emanuele Severino. and the Grand~~
~~Inquisitor. and various acts of excommunication, in~~
~~furtherance of truth, through acts of censorship and~~
~~"justified intimidation". if I can finish it, this poem~~
~~will be a poem about Galileo, in Rome, where the~~
~~Sun revolves around the Earth. if I can finish it, this~~
~~poem will be a poem about George M. Johnson~~
~~and Jina Amini. if I can finish it, this poem will be a~~
~~poem about Rashida Tlaib. and Ilhan Omar. and the~~
~~fear of the Other empowered to speak. if I can finish~~
~~it, this poem will be a poem you'll pass around.~~
~~I'll keep writing it over and over until I get it right~~
~~because, *any golfer who has watched his partner*~~
~~*shank a short approach knows, it would be absurd*~~
~~*to accept the suggestion that the resultant four-let-*~~
~~*ter word uttered on the golf course describes sex or*~~
~~*excrement and is therefore indecent... It is ironic, to*~~
~~*say the least, that while the FCC patrols the airwaves*~~
~~*for words that have a tenuous relationship with sex*~~
~~*or excrement, commercials broadcast during prime-*~~
~~*time hours frequently ask viewers whether they are*~~
~~*battling erectile dysfunction or are having trouble*~~

going to the bathroom. if I can finish it, this poem will
be a poem about a poem. it will be a poem about
the shaping of the shape of truth. with redaction
and silence. to the dismay of some, Rachel Corrie,
Nikole Hannah-Jones, and Shireen Abu Akleh will
appear in the lines inside this poem. and it will be a
poem. it will be a poem that someone wants edited.

[the Bureaucrat]

active oversight processes, including regular
and ongoing monitoring, are desirable. effective
systems of internal control provide the basic foun-
dation upon which the structure is administered.
these systems of internal control must be routinely
and continuously evaluated, and, where necessary,
supplemented with additional management prac-
tices so that weaknesses are promptly corrected. as
such, reports are necessary. reports are required of
all levels of management. this will facilitate a contin-
ual ongoing assessment and strengthening of sys-
tems of internal control consistent with appropriate
chain-of-command reporting relationships as well
as any number of relevant stabilization procedures.
this will enable, as appropriate, a modification
of integrated operations. if the recommended
reporting function is standardized, there will be
additional opportunities for course correction.
this, in turn, will help resolve the findings of audits
and other reviews and will enable all elements of
a satisfactory system to be present through the

ongoing forward motion of regulated inertia. the further development of policies and procedures adequate for compliance with applicable criteria, standards, and other requirements is also continuously necessary. simply put: follow the standards; engage in a routine application of internal controls; avail yourself of opportunities for correction; and note any corrections through the standard reporting function. all systems and processes will ensure that ongoing activity carries forward compliance requirements in a manner consistent with sustained, systemic, risk-mitigating quality control.

[the Prophet]

behold I saw empty space and immeasurable dis-
tance. and behold I saw a tiny prick of light. and I
saw this light expand. and then I saw it gain form
and weight and mass. and I saw animals issue forth
from inside a sphere inside the massing space. and
then I saw men and women walking. and the men
and women were groping and feeling about the
air. they were interrogating time with their hands
and mouths and fingers. and they were poking and
gesturing and pointing at something unseen. and
as they moved, they murmured and muttered. they
licked and smacked their lips. the men and women
thumped their chests. they stroked the fur of the hair
of their beautiful beards. and they decorated their
bodies with shells and cords and leather-strung
ornaments. and the people wrote all over them-
selves. they covered their bodies with symbols and
pigments. they smeared clouds of color on their
closed-open eyelids. they tattooed their body parts
with needles and feathers and long pointy fingers.
and they taught themselves to count. they learned

to add and subtract. they learned to multiply. the shapeless shape of meaning. and so they made up stories. and the stories became other stories. stories within stories. stories that grew in and out of other stories. and all these stories metastasized. and the stories penetrated one another such that each story always-already referred to some other story. and the words of their narratives comingled. they grew into a thick film that was drawn across the land such that no one could see or even remember anything that resembled the unwritten present. and it was then then that I saw billowing clouds issue forth from the crowded density of words. and I saw smoke. and the sky going black. and the land over-run with fog and darkness. and behold the world was a bloated mass of bloated masses. and the Earth fell through space like a deflated ball going limp in a weak trajectory. and as I looked on, I could see the people, at times, crawling in and out of the useless ligatures of ever-growing, ever-longer sentences. and the people were slithering about. and as they crawled about, they bent and pulled, and tugged at, the levers and buttons of symbolic meaning. in this manner, they flicked each other's switches. they stroked and turned and spun one another's dials. and the men and women's tongues, they slipped in and out of their mouths. they crept forth from their open and shut gaping red mouth holes. and these people said *yum* and *mmmm*. they said *hmmmm* and *wow*. and as I looked on I noticed (yet again) more (billowing) great fires. and there was extreme heat and cold. and I saw pestilence and knifing white jagged blue lightning. large and small stars were falling from the skies. and birds and satellites were falling from the skies. and the people, they called out. they whimpered. they cried. they cried and yelled and pointed at one another. and the people babbled in high pitched frequencies. they droned. and monologued. and as I watched, I saw large numbers of animals begin to disappear. frogs and toads and salamanders disappeared. and numerous species of birds. and bees stopped buzzing. and I saw metal scorpions multiply over the cracked earth where no plant would grow. and everywhere, as I looked, I saw bit-ing flies swarm and feast on the sleepless countless dead. and then the amoebas awoke. they crawled forth from the earth where the permafrost melted. and there were masks. and bubble suits. and all around, I saw masses of scurrying numbered black and white vermin. and the seas began to boil. and black goop bubbled from the ground. it ran into the rivers. it ran into the lakes and oceans. it collected in mud puddles and cereal bowls and styrofoam coffee cups. and behold I heard flutes and drums and the blare of bugles. I heard high sounding cymbals. and then I took off my glasses, and took off my tie. I took off my lab coat. I began to chant. I began to gesticulate. and dance. I danced cursing both the earth and sky for all I had witnessed. and so, I slurred amid the stench of my own foul smell-ing vision. and I spoke out loud. I spoke out loud with a blue swollen tongue. I spoke, then and there, through locust-stained teeth. and I said, *there will be more*. and there was more. and more. and the Earth shook. and wept. and the wild tamed people of the Earth fell down and fidgeted on the cracked and broken ground, now scripted over with too much (and too little) meaning.

After the last monologue has been performed the player (dressed as the Prophet) turns out the light one final time.

The audience hears the door open and close.

The light again slowly comes on revealing the absence of the player as well as the presence of an empty chair in what is now a fully illuminated empty white room.

The light in the room slowly dims and goes out.

[two notes to to be read
and studied aloud]

Another stage set. A bench in a park surrounded by trees. In the background, the silhouettes of sky-scrapers. Two players dressed in black wearing Saru masks are seated on the bench. Player One rises from the bench and reads Note 1.

Note 1

your words, or maybe mine, could maybe make the sky go silent. could perhaps make the mind unravel. amid the ragged fictions of the self, my days are spent. and I am used. I am the used plaything of self, exposing each other me as some chance fac-totum to my body (and all it reaches for). my body reaches for the cool liquid of dry scratched-out and thumbed-through meaning. for the jagged innuendo and bleeding type of typeface hiding or residing in some stammering feral vernacular, a vernacular composed in and with purposeless words. these cluster amid all my tiny lyric minutes, the unwashed slick and dirty moments that fabricate our days like hands. and mouths. and fingers. these fingers full of sand.

Player One bows and sits down. Then Player Two stands up and reads Note 2.

Note 2

the way the world is arranged: will without reason. or reason as an affect of will. a will without object or destination. like the way my mouth and hands are arranged. in memories of measureless trajectories of lexical reference marked with soft punctuation. an episodic soft punctuation of repeat iterated pleasure such that I can feel the meaning of it, the meaning of some present distant literal reference, some unknown hoped-for imaginary locus. it is in this selfsame locus where we will discover the freedom of words (again).

*Player Two bows and sits down. Slowly the theatre
lights dim and go out.*

[preliminary notes on
the status and position of a
figure pasted to a landscape]

cross it out. erase everything. paint over it. redact
it with a black magic marker. now scribble over the
script until the text is obscured below broad strokes
of black redaction marks. these signify nothing
except meaning removed.

now abandon all linear sense of movement, over
and across a number line. cut through it laterally
and move outward in a series of arcs or circles.
now, for the simple sake of it, walk around and drift
above your feet. float freely and *come into space*.

come into space says Charles Olson. with your four
limbs unfolded. with your sore days given. to lists
of arbitrary death. to lists of routine cause or casual
consequence. because the streets are demanding
it. are asking you for the shape of your meaning,
here, in this moment, in these open spaces of

wanting, and needing, and having.

dis-adapted and malformed, we come to fashion our lives by writing them down. we insert words in the intersections and maybe-spaces we cannot explain so that we might render meaning out of nothingness. *without music, the world is invisible.*

[*Untitled (1999)*]

Untitled (1999), a work of sculpture by Anish Kapoor, is a large nine foot lacquered stainless steel concave circle. it is dark and it is blue. from where it sits in the Denver Art Museum the dark blue steel circle refracts light in a particularly arresting manner. after the light is captured, it is reflected outward in a display of electric blue sinusoidal waves that jaggedly play over the radii of the sculpture's darker metal surface, so that, as one approaches the sculpture, the refracted light appears to move and spin across the sculpture's concave rounded hollows. as if *Untitled (1999)* were some kind of midnight-blue-black wormhole. or else some type of cosmic vinyl playing the recorded sounds of an empty silent universe. in this manner *Untitled (1999)* appears to be some sort of interstellar portal. one that was designed to facilitate and produce *spooky action at a distance*. and yet *Untitled (1999)* is also form and weight in space. and it is also energy masquerading as form, so that, echoing the curator of the Denver Art Museum, we might speculate that *Untitled*

(1999) represents the void or abyss. or maybe *Untitled (1999)* represents some unseen world far removed from all human experiential access. as such *Untitled (1999)* also represents human spec-ulation about what it does not see or feel. in this way *Untitled (1999)* always-already implies a desire to both touch and feel the unknown, that which is ever out of reach.

[the Ear poem]

this space is reserved for the Ear poem, a poem
about the act of listening. this space is reserved
for the Ear poem. an ear is an ear. it presupposes
other ears. this space is reserved for the Ear poem.
the Ear is a vessel floating in sound. it exists in a
field of play. like a boat. like a canoe on a river. or a
flower in a field. and so the Ear that appears inside
the ear poem, situated now, in a field of other Ears,
is nestled in the presence of both time and space.
yet the Ear poem is also stationed in history, a place
where language unfolds, like a flower that blooms
in a process of becoming-unbecoming, a place of
change, a locus of stasis and motion. and so the Ear
poem vibrates. and the ears inside the Ear poem
vibrate. they quiver and they shake. they dance and
they shiver. the Ear poem is attached to a book the
way an ear is attached to a person, the way a per-
son is attached to other people, who are, in turn,
attached to other people, people who wait in a field
(a field of fields) of situated meaning. one two three
four. in the Ear poem people repeat. in the Ear

poem people repeat just as words repeat in phrases or paragraphs where meaning expands. where it comes to be arranged as a kind of relational sediment. that is to say language accumulates. it gathers. it amasses. it piles up upon itself in the *longue durée*. and so the emergent body of language grows where words bend as they curve (recursively) in the place of a space reserved for an Ear poem. inside the dis-adaptive authoritative dispersed geography of the Ear poem, the ears of very many people exist. these ears live in the East and West of here and now. they live in the North and South of there and then. and they live and they breathe. and they speak and they listen even as the content of the meanings transacted comes to be saturated. and sometimes the ears of the Ear poem exist in the arranged spaces of living breathing buildings. sometimes they exist in the lengths of hallways and passageways. ears move across thresholds. they move over borders, passing, as they do, through gateways and doorways. and still this space is a place reserved for an Ear poem. in the Ear poem ears appear to peer through windows. yes yes. I am speaking of ears climbing up mountain sides. one two three four. I am speaking of ears that move in and out of the vibratory locations I will now refer to here as "interlocutory space." five six seven eight. and still this place is the black and white space reserved for an Ear poem. the words of the Ear poem reserve and organize space with a coded, stratified diction. as such the Ear poem operates the way a society operates. one two three four. there is a certain hierarchy of meaning. yes yes. and there is contested space. and there are instabilities. and geographic topologies of power. there are centers and nodes and clusters of motion wherein the force of meaning moves to the beat and pulse of papery stillness. and still this is the place reserved for an Ear poem. a place where the lines bend and turn to disclose meaning to all the assembled ears that murmur amid the playful music of time. do you hear it? are you here with us? are you listening? the Ear poem is a metonymic machine devised to engineer further discussion about the emergence of mean-ing from matter. and there are many different ears here. they come from the four corners of the Earth. they arrive at variable speeds. they arrive at both opportune and inopportune moments. inside the Ear poem the potential field of people who could now be listening is exceedingly vast. that is to say

the field of potential ears numbers well into the
billions. I'm guessing that maybe there are 16 bil-
lion ears who might could maybe listen. one two
three four. in the space of the Ear poem a place has
been reserved for you. five six seven eight. will you
sit down? will you listen to the sounds, the auditory
sensations, the well intended, disclosed, quivering
locutions binding you to what has been said inside
the space of the Ear poem? this space is a place
with and without meaning. it is a place with and
without feeling. we don't even know why. perhaps
we don't care.

[how it is]

you are screwing in the top of your head. you are placing your hands all over the uneven surface of some hollowed-out body. you seem to want to. you seem to need to. to need to want to. to want to seem to. to want to find a way forward. into this minute.

[how it is]

diagram of night. in the upper right hand corner, the soul with its pencil is scribbling. is trying to find: a place to hide. a place to peek out at the world. at all that is lost. and all that is stolen.

[how it is]

you put on a hat. you put on another. you put on
a third and stand in smoke. nothing you worship
appeals to you now. nothing you pray to will speak
through you now. you are alone. and given. you
are withdrawn, but also, you are taken away. to be
delivered. to be pressed forward into the pressed
pages of a book. some literary quantity of quality.

[meta and qua]

a long and narrow room in a museum. the room
has a low ceiling and slants away diagonally to the
left. it is dimly lit and at the back, in the center of
the farthest wall, there is a barely discernible door.
amid the darkness of the room, one sees five people
standing in shadow. each is positioned equidistant
from the most adjacent person and from the wall
behind each and all of them. they collectively ap-
pear to form a line slanting away from the entrance.

lighting is arranged in the room so that the farther
one is from the entrance, the darker it gets.

to the left of each figure, a standing lamp. in front
and to the right of each figure, a microphone.

each figure is positioned so that they can easily step
to the nearest microphone.

Speaker 1

*Speaker 1 is a woman dressed as a professor or intel-
lectual. when the performance begins she steps to
the microphone and the light of the standing lamp
adjacent to her illuminates. she then withdraws a
stick of lipstick from her pocket and applies it.*

what an instrument is
lipstick theater. I, speaking of
the writer's power to evoke,
suggest to you, the reader, through
telekinesis and floral patterns of
looped melodies, a motif, a
watermark for ensembles such that
through diction, and by means of
jumped-up syllables, I will give you
the fragments of a story made by means
of loose contingency and broken
circumstance. this story is perpetually
becoming out of nothing. it is
written both inside and against

the stammer of our flesh.

when Speaker 1 is finished she takes a bow and steps back to the space she occupied before. the light of the adjacent lamp winks out.

Speaker 2

Speaker 2 is a man wearing eyeglasses.

Speaker 2 steps to the microphone and the light of the lamp adjacent to Speaker 2 flashes on. Speaker 2 withdraws a cloth from his pocket and uses it to clean his eyeglasses before addressing the audience.

in liquid night a storyboard appears in order to wake you. it references the dreams you trace across the holes that rest atop your tiny tiny eye parts. perhaps these dreams contain scenes built with covert abstract plot lines. and perhaps these plot lines are intended for an experimental purpose. as such Beckett appears in order to wake you, dragging with him the hidden imperatives of vanguard drama. and so you prepare yourself for an absurd fate. you dress yourself for all creation. you get down on all fours in front of a microphone. and you begin to growl. you bark. you bark and you growl like some small god in search of a limit. and maybe you will find this limit. maybe you will become this small god's tiny mouthless counterpart. or maybe you'll become bigger and larger. at this point, in the storyboard sequence placed inside this scene you have become large but also teeny tiny. you are itsy-bitsy. you are micro or nano, or huge, or even bigger, or larger as you undress and return to all fours, compelling yourself to growl and bark completely *desnudo*. and so you are naked. and ghostly. and feral. you are painted in print, a thinking, speaking, barking dog whose every bark and growl alludes to the connection between nature and culture using little more than the animal sounds you have now pushed inside the microphone.

Speaker 2 gets down on all fours and barks and growls. Speaker 2 stands back up and takes a bow. Speaker 2 steps back to the space he previously occupied. the light of the lamp goes out.

Speaker 3

Speaker 3 is a short man in a pin striped jacket and torn demin jeans wearing a clown face and a red bulbous clown nose. light comes on after he steps to the microphone. he speaks slowly, then rapidly, and then again, slowly.

perhaps my vision for this piece is always-already borrowed or stolen, regardless of the way I experience life. and perhaps language is a garden of white flowering vowels.

at this point Speaker 4 seemingly pulls a bouquet of white flowers from the air.

or maybe I could describe the flora and fauna of words using stylized phrasing that is somewhat more concrete, more given over to the visual. for instance, the sad Moon being tied down and felt up with black leather gloves.

the clown man now holds up his hands to emphasize

that he is wearing black leather gloves.

here I wave to you from inside the grey rubble of a static life. and in this life, in this performance, in the image now seated and unbuttoned before you, a bad omen in torn jeans. it bleeds out adagios and chorus songs that no one sings or reads.

now a chair descends from the ceiling. the clown sits in the chair for thirty seconds. he stands. the chair ascends to the ceiling. the clown moonwalks back to the space he previously occupied. he waves at the audience with his black leather gloves. the lightbulb of the clown's lamp flickers before going out.

Speaker 4

the appearance of Speaker 4 is left to the imagination of the reader (or if this piece is actually performed, Speaker 4's appearance is left to the discretion of the director). the light of the lamp goes on. Speaker 4 steps to the microphone.

maybe language is about the presence of what is removed or gone (what is negated or absent). or perhaps language is the rarefied power to create that which isn't, using references, like animal tracks, to that which is (or was). and so lists of stolen (appropriated) content. or mathematical equations. or other kinds of referential symbols (like spiral swirls inside of spiral whorls inside of time-worn labyrinths).

in this episode Heidegger is fleeing from Beckett. that is to say the absurd is chasing the authentic across the town square and down the street. and there is a black cab idling in background noise. and there is a mass of limbs, a seeming clump of bodies. these bob and shake and vibrate in and out of view until they stir the nerves to move the situated self to experience the upper limit music of the flesh. simultaneously, all around, testaments of, or, for, empty rooms in flight.

picture a process of dissolution coupled to the abandonment of reason. what Goya called the sleep of reason. the older I get the more I imagine it. the abandonment of purpose coupled to the drift of social process. a somehow collective inertia signaling our soon-to-be mutual and collective unraveling. it's a kind of simultaneous spreading out and creeping forward: into accident and entropy.

and so I respond with words and phrases and ripped-up sentences. to wit, I respond with meta-poetic messages concerning the bloom of symbols and the flower of meaning. in this manner I draw massive tiny tiny diagrams. in this regard small pictures of time's far-flung wonder. doodles of desire plugged into desire and desire such that towards each and all my ends I push and pull the levers and buttons of an imagined human heart. this human-primate heart beats with and without purpose. it gives form to "truth" where "truth" is merely the ongoing, periphrastic, self-defining, unbroken yet unfinished, fissuring, disjunctive construction of something from nothing.

Speaker 4 takes a bow and steps back to the space he or she occupied before approaching the micro-phone. the light goes out.

Speaker 5

everything vibrates. even things that don't appear to vibrate are always-already vibrating. and sometimes the things that are vibrating beat their beat inside a heart I keep inside a tiny tiny suitcase. this tiny suitcase resides inside all or part of me. and still my mind like a sign that says no. that whispers no passing zone. or else my mind like a sign in the center of a black square that occupies the big void, the big zero, of emptiness, and well, sort of, *what if.* what if the Sun is a woman and the Moon is a man? what if between the two of them they make a certain sort of eclipse. perhaps this eclipse fuels a poetry of circuits. perhaps it feeds a poetry of futile resistance. because I am willing this setting and scenery into existence using little more than the clutter of my mind. my mind that trembles like a sign that still says no, no passing zone, but maybe…

maybe and, *what-if?*

when Speaker 5 is finished she takes a bow and steps back to the space occupied before approaching the microphone. the light goes out.

after all the speakers have read their monologues
the room lights up as the performers undress. the
performers strip off their clothes while staring at the
audience. now five wrapped packages appear over-
head attached to five ropes. the ropes and packages
descend from the ceiling until the packages reach a
distance approximately four feet from the floor. the
performers remove the packages and open them,
revealing five black folded jumpsuits.

the performers unfold the jumpsuits. they pull them
on. they pull the pants, legs, and sleeves over their
arms and legs. they pull the jumpsuits over their
torsos and zip them up.

now a second set of ropes descend from the ceiling.
attached to each of these is a Kyōgen Saru monkey
mask. the masks slowly descend from the ceiling
until they are five feet from the floor.

the performers remove the masks from the ropes

and place them on their faces as they turn to face the audience. the performers wave and bow in unison. they bow again in unison. the performers turn and face the back wall. a door in the center of the back wall opens and the performers collectively walk in a line towards the door until each performer files out. after all the performers have left, the room goes dark.

[photo of the no-face of meaning]

the ape liked the idea of the book better than the book. the ape liked the idea of the body better than the body. the ape liked a sense of repetition better than the actual acting out of all the iterated instances the ape often would return to.

you see the ape was in love with ideas. and form. and concepts. and the ideas (and forms (and concepts)) he (or she) kicked around her head became, in a certain sense, the thoughts that made him or her a certain kind of animal, one that moved through the day using words to accompany his or her (or their) body and what to do with it.

and so ideas (and words) taught the ape how to give form to matter. and this mattered. this matter of matter becoming references to matter with words that moved in and through the unconscious of the ape's plugged-in living body.

and so he or she or it or them, well, they built

themselves a set of selves. and they lived inside these selves, these roles they scribbled down on paper.

besides, he or she (or it) (or them) (the apes that is)—they were laying themselves down and standing themselves up (inside of time) (inside a place) (a spherical space) constructed outward and forward in an ongoing expansion of intensity that pushed the ape or apes forward toward limitless extension.

that is to say, the ape or apes, had descended from trees with the inside-out of immanence. and their minds became (over time) living machines emergent from matter. and these machines operated on them just as they, in turn, operated on the machines the apes now used to move their bodies in and through the landscape.

you see the apes, they were being pushed here and there. and they were pulled along. and they were pushed and pulled and pushed and pulled and they were pushed and pulled along. by all the words they had written down in sentences. and phrases. and paragraphs. that heralded their coming incandescence.

[*language and death*]

what is. this is. that is. this here. this this here. this being-here. and the memory of it. and the suppositions that grow from it. all the wandering-wondering (that follows from it). all the pointing to and gesturing toward. that which stretches forth. as immanence. the emanations, in turn, becoming the being of becoming, a becoming-being unfolding amid each and all the nameless coordinates we try to bind and yoke. with names.

and so a looking at. a peering into. the *this* of the it of the it of *being*. the continual growing shape of it. and its perplexing uncanniness. the strange peculiarity that inheres and abounds everywhere. that suffuses the total inexplicable *this of now*. and the way our experience of it is coupled to an anxious realization that our life is conjoined to, and opens onto, an eventual unavoidable end. the end of me. and you. and us.

we stand here, totems of flesh-time, organic living obelisks emergent from matter, erected in time-space. we stand amid and before the this of this. and we shuffle our feet. and we say *umm* and *hmm*. we say oh and well. we mumble in our socks, standing lost and befuddled in our worn out shoes as we confront the unspeakable what of now.

and so there is a need to know *why*. there is a need for meaning, such that a kind of gilded linguistic latticework begins to grow all around the *this-here* of our experience of *what is*. and the framing that emerges is a kind of dis-adaptive response to the immediacy of our presence sutured to the known and felt certitude of our own finitude.

as such language becomes a set of stage directions for dealing with the empty nothingness of now. perhaps the stage set is the empty set, the null set. and maybe each person, as actor and player, acts and speaks from the knowledge that he or she is a *placeholder of nothing*.

with language we stand and confront our own

becoming unbecoming. we use signs and symbols to work meaning into the liminal spaces of our lives, even as our lives continue to open themselves onto all the intersections of *here* and *not* here.

[the GOD poem]

this space is reserved for the "God" poem. it is a box. it is a place to put God inside a box. but God is unsayable. this space is reserved for the "God" poem. yet God is forever unsayable. this space is reserved for the "God" poem. and yet God is invisible. this space is reserved for the divine concept of "God," a supernatural heavenly being resplendent and glowing like silver (or gold). God now sits here, in this discursive box, and says to himself, "though I am very very shiny I am still unsayable, I am unknowable." God is unsayable. God is unknowable. this space is reserved for the "God" poem. inside of it, God says, "I am. I am the great I am." God who is the Alpha and Omega, is invisible, unknowable, and unsayable. this space is reserved for the "God" poem. God is unsayable. this space is reserved for the "God" poem. yet God is unsayable. and unknowable. God is energy. and God is heat. God is light. God is time. God is space. and yet God is unsayable, unknowable, and always-already, God is completely invisible. God, the great "I am" of the invisible whatever-hope is, perhaps or maybe, something out of nothing. and so God is the beginning and the end. God, in his unsayable, unknowable godlike godly essence, says to you and me, "I cannot be contained," which really means that God is something like a transcendental signifier. and so God cannot be contained. by words. or by language. yet God inheres in words and in language, being the presence of an absence that always-already implicitly grounds all meaning such that God is actually the hidden present name of the unreachable "Real." put differently, God is the dis-adaptive rock bottom premise of language. and yet the essential premise of God is that language cannot ever actually speak or capture or contain the power and the glory, the divine electricity, the dark matter, the blood red ribbons and falling black confetti, the glowing prismatic spray, the color and the lights, that are/is "God." because God is unspeakable, unknowable, and invisible. that is to say, God is invisible yet always-already present in the unseen interstices of the flowering vowels and consonants emanating like energy from the material "Real" contingent not-for-us, a cosmic space of play where God is simply: *the Idea*, emergent from matter.

emptiness. yes yes. I tried to write a book about emptiness. one two three four. but I could not write a book about emptiness. five six seven eight. so I tried to organize my notes in a *nota bene* format. yet I could not organize my notes in a *nota bene* format. the texts I'd thought I'd written somehow now escaped me. all the vowels, each and every consonant, winked and blinked and fled, moody, brooding, backwards, into the open spaces now open all around me. given this context, given the situation, and no intersection with logic, I made up new meanings. with the stubs of my fingers I wrote a new sentence. this sentence described the strange appearance of this very same sentence. it became somehow now, a growing narrative of some untold different story. a story I might have, could have, finished. but I knew now that the purpose was ever not to finish. the purpose is never now to finish. so

I called forth fragments. and multiples. from time and from space I brought forth energy. and matter. in this manner I referenced rocks and clocks and planets and galaxies. to wit I spoke of consonants and continents. I spoke of sand. I referenced water. I held forth at length about oceans and rivers and tectonic seabeds. and all along I pointed out the flames. the billowing great fires. and instances of ongoing recurrent cataclysm. I pointed to the gleam and flash of teeth. and the many different names of diverse dead and living animals. I mentioned, for example, strange forgotten camels. also unknown unseen temnospondyli radiating outward to inhabit aquatic and subaquatic habitats. I used each and all of these to source the source of language in processes of biochemical physical kinetic immanence. and so I laid forth the physics of why and the story of the birth of wonder, a multipart serial recitation speaking to the becoming unbecoming of meaning from matter. perhaps this story is now unbound forever more, for never ever after. because the logos of thought is subject to lack. and it is animated by a certain sort of excess. this excess vibrates out of passages plugged into "the Real" even as "the Real" itself gives rise to the represented unreal using an alchemy of words. these *live and move and have their being.* in the dream of a perfect paragraph. in a variable sequence of mellifluous sound composed at the margins of why. or wherefore. or *wherefore art thou.*

[notes on the possession
of animals by spirits]

this is not a fable. nor an allegory. but once I saw the layered edifices of language climb out of a set of carnally connected bodies. these bodies were positioned in time-space. were seated and arranged amid the flowering petals, the pistils and stamens, of affect and energy. and the bodies' arms and legs were there. and their hands and feet were there. were folded beneath and in front of them. and the bodies' heads were there too. were material crowns of presence seated inside and amid a series of repeat humming intonations vibrating themselves alive in some present sequence of saying.

in this context I thought about all the machinations of the heart, by which I mean the grasping clinging possession of the human heart by an untamable (yet rational) mind, a mind which speaks and bleeds into (and out of) the four valves of the heart, the heart and mind both somehow now (and always) being fed and powered by two lungs, these two lungs concurrently plugged and patched into a network of circulation rooted to the vibrant tonality of the Earth's damp wet soil (and all its cool clean air).

still, for some reason, there was a left hand and a right hand. there was a world inside a void inside a world inside a void and all the numbers and letters and words were arranged to describe the real and actual basis of the becoming-contingency of belief and the way that it seemingly appears to germinate and flow forth from ever multiplying fields of materially immanent impermanence.

still my inadequate articulation remained imperfectly complete. as if measurement and language existed and floated at the back of time, on a celestial clock, the clock itself appearing without any sort of discernible clock face. the clock was just some no-face affixed to a wall in a cosmic endless hall without windows, the walls of the hallway shifting even as the floor and ceiling disappeared inside themselves such that the walls refused to house us (or anything at all).

and so we lived, for a time, with no meaning. and in this place (of no meaning) we made up our own meanings. meanings without any underlying purpose (to the meanings).

this may be vague, (or confusing), or perhaps even nonsensical so let me attempt some straightforward clarity. there is no wherefore. there is no whomsoever. for it is written: *there is nothing to say now.*

[the last dance]

1

you climb down a ladder and descend into a room.
the room is a glass cube encased in darkness. in the
distance you see a shimmering light. it hangs there.
it shimmers, fixed amid the stillness.

2

now your eyes adjust to the light and you see a
lit-up rock where the light once was. the rock is
suspended above the floor of the glass cube room
by a steel rod, and though the purpose of the rock
is unknown you are certain the rod was designed to
hold the rock in place, here, in this room, heavy and
laden with the shadow of meaning.

3

the rock stares at you. it reads your thoughts. it communicates with you telepathically and beams into the round head of your mind, thoughts which speak out of turn. the rock then becomes a small yellow dot. it is represented on the page as a small yellow dot. this dot inhabits some lines inside a poem. you don't know why. until you do.

[in the clearing, a final curtain]
(draw it yourself)

now a black curtain falls across the page. and now a black wall slides down in front of the curtain. and even now a line descends from the center of the top of the page and slowly forms a sigil or a glyph. it takes shape as rectangular form, as a squarish long frame, the axis vertically oriented from top to bottom, as if it were some sort of door or window. yet now another line forms to bisect this frame. it forms approximately midway between the top and bottom. the line might be said to signify the horizon. and now, even as you watch, a semi-circle affixes itself to this horizon, the two ends of the semi-circle attaching themselves to the approximate space where the horizon line intersects the two vertical sides of the rectangular frame so that the visual appearance of the semi-circle and its placement suggest an object that is partially visible but also partially invisible, the base of the object resting somewhere in the space beyond the horizon. and

now midway between horizon and the lower out-
side bottom of the rectangular frame another line
forms. this line also stretches from the left side of
the frame to the right side of the frame, running
parallel to, but also beneath, the horizon line. this
secondary line perhaps signifies the ground (inside
the frame) inside the mind (inside of time). yet also
now, outside the contents of the frame, in the space
right below the frame's bottom-ended border,
another line takes shape in space. this newer line
finds its place amid (but also below) the made
arrangement of all the other pictographic elements.
it takes form as a wavering shape, as a sort of bent
and vibratory squiggle, the features of it perhaps
alluding to both vibration and energy. this wavy
line, having been placed outside the main frame,
might also represent that which is not only unseen,
but also, that which is unknown (and maybe even
unknowable). because we do not ever know what
we cannot ever say.

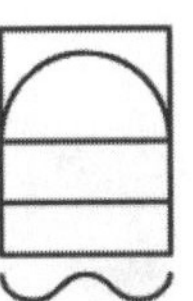

NOTES and ACKNOWLEDGMENTS

1. Notes on *notes*

Notes on the possession of animals by spirits (hereafter, *notes*) offers a speculative geology (and genealogy) of language that roots human meaning-making activity in a broader set of cosmological, geophysical, biochemical, and historical processes. The books *Philosophies of Nature After Schelling* by Iain Hamilton Grant, *After Finitude* by Quentin Meillassoux, and *Nihil Unbound* by Ray Brassier provide much of the inspirational impetus for the book's ideational content, animating its central questions and motivating the overall trajectory of its developmental arc. Jaques Lacan, Martin Heidegger, and Giorgio Agamben figure prominently in the book's texts addressing language's emergence from and relationship to human practical (and impractical) activity. Keiji Nishitani inspires the book's zen-infused affirmative nihilism while Julia Kristeva's thought manifests itself in the book's treatment of poetry as a super-charged conduit for nature's vital energy. The enduring influence of Gilles Deleuze, a related affinity for Elizabeth Grosz's philosophical works on ontology and art, as well as an appreciation for Pierre Joris's *nomad poetics* should also be noted.

2. Sources, References, and Allusions

The text "[sketch for an installation]" is inspired by three sources: 1) early Babylonian and Sumerian writing on stone pillars; 2) the standing stones, cairns, and neolithic ruins of Scotland, Ireland, England, Wales, and France, and 3) the monolith in Stanley Kubrick's *2001: A Space Odyssey.*

The brief dialogue and dramatic sequence, "[a dark room and two voices]" alludes to Plato's "Allegory of the Cave" and Samuel Beckett's *Waiting for Godot.*

The serial poem "[three scenes for a theatre of one]" is a textual response to Iain Hamilton Grant's *Philosophies of Nature After Schelling.* The titles for this sequence are inspired by the cinematographer Kathy Kasic's installation, *ENTER THE WIND,* which she characterizes as "cinema-for-one".

The poems "[psalm of the not-for-us]," "[something from nothing]", and "[the book of geologic time]" are inspired by and allude to ideas, processes, and concepts found in Quentin Meillassoux's *After Finitude,* Ray Brassier's *Nihil Unbound,* and Iain Hamilton Grant's *Philosophies of Nature After Schelling.* These texts also find inspiration in the writings of Gilles Deleuze and the "inhumanist" poetry of Robinson Jeffers. Earlier versions of [psalm of the not-for-us]," and "[something from nothing]" appeared in *The American River Review* No. 30.

Several sources were consulted while writing the hybrid poem/creative-non-fiction text/performance piece "[a book of geologic time]." These include *A Concise Geologic Time Scale, 2016,* by James G. and Gabbi M. Ogg, the Smithsonian National

Museum of Natural History's web page (https://humanorigins.si.edu/evidence/human-fossils/species), various UC Berkeley webpages on Geologic Time (https://ucmp.berkeley.edu/help/timeform.php), and several Wikipedia pages accessed through a gateway page available at https://en.wikipedia.org/wiki/Geologic_time_scale.

The spiral symbols in part VI of "[a book of geologic time]" were rendered by C.J. Rosenthal. The diagrams on page 41 are based on spirals carved into rocks and stones found at Newgrange, Ilkley Moor, and Nine Mile Canyon.

The poem "[descending the tree of life]," and the serial poem/installation text "[a field guide to bipeds]" are inspired by Stanley Kubrick's *2001: A Space Odyssey* and Miroslav Holub's "Hominization."

The serial poem/installation text "[a field guide to bipeds]" draws on research contained in *The Human Past: World Prehistory & the Development of Human Societies*, edited by Chris Scarre, *Before the Dawn: Recovering the Lost History of Our Ancestors* by Nicholas Wade, *Processes In Human Evolution: The journey from early hominins to Neanderthals and modern humans*, edited by Francisco Ayala and Camilo J. Cela-Conde, and *Sapiens: a Brief History of Humankind* by Yuval N. Harari. Websites consulted for additional research include the Smithsonian National Museum of Natural History web page (https://humanorigins.si.edu/evidence/human-fossils/species) and each of the Wikipedia pages providing general information about each of the hominid species featured in these poems.

The title of the narrative serial poem "[being and time]" is taken from the title of Heidegger's canonical 20th century philosophic work on ontology. This poem uses Heidegger's work as well as ideas found in works by Agamben and Lacan to imagine and conceptualize the processes that led to the emergence of language. Graham Harman's *Heidegger Explained: From Phenomenon to Thing* and Daniel Everett's *How Language Began: The Story of Humanity's Greatest Invention* were also consulted.

Sources on human history and prehistory used to partially develop the content of "[being and time]" include *The Human Past: World Prehistory & the Development of Human Societies*, edited by Chris Scarre, *Before the Dawn: Recovering the Lost History of Our Ancestors* by Nicholas Wade, and *Processes In Human Evolution: The journey from early hominins to Neanderthals and modern humans*, edited by Francisco Ayala and Camilo J. Cela-Conde.

Section III of "[being and time]" refers to the discovery that *Homo erectus* very likely had the capacity for symbolic representation. See the article "Zigzags on a shell from Java are the oldest human engravings," by Helen Thompson, December 3, 2014, *Smithsonian Magazine*. Also see related content in Daniel Everett's *How Language Began: The Story of Humanity's Greatest Invention*. Everett's ideas about the linkages between tools and language-use are not dissimilar to Heidegger's ideas associating "ready to hand" tool-use with the human process of "worlding" through the use of words.

Section IV was developed after reading about cultural artifacts discovered in the Blombos and Wonderwerk Caves of Africa. See research in *The Human Past: World Prehistory & the Development of Human Societies* edited by Chris Scarre, as well as the article, "Early Evidence of Brilliant Ritualized Display: Specularite Use in the Northern Cape (South Africa) between ~500 and ~300 Ka," by Ian Watts, Michael Chazan, and Jayne Wilkins, in *Current Anthropology* Vol. 57, No. 3, June 2016.

Section V and Section VI of [being and time] address the appearance of Venus figurines and cave art in the archaeological record. Clayton Eshelman's poetry is an obvious and important forerunner. Consulted references include a Wikipedia page on Venus figurines (https://en.wikipedia.org/wiki/Venus_figurines) and the following books: *Prehistoric Art: The Mythical Birth of Humanity* by Jean-Pierre Mohen; *The Nature of Paleolithic Art* by R. Dale Guthrie; and *Images of the Ice Age* by Paul G. Bahn.

Section VI of the poem "[being and time]" alludes (again) to Plato's "Allegory of the Cave".

Section VII of the poem "[being and time]" explores the development and use of written language in the Neolithic period and early Antiquity. References consulted include *The History and Power of Writing* by Henri-Jean Martin, *The Gods and Symbols of Ancient Egypt* by Manfred Lurker, *Inside the Neolithic Mind* by David Lewis-Williams and David Pearce, and *The Book Before Printing: Ancient, Medieval, and Oriental* by David Diringer. Other sources include Wikipedia pages on the history of written language. See for example, https://en.wikipedia.org/wiki/History_of_writing.

The poem/performance text [Prospero speaking amid the residue of time facilitates a bridge to the present] features the Shakespearean character Prospero. This piece references a series of literary epics (and/or their main characters), alludes to the scholarly work of Henri-Jean Martin, and draws directly from ideas and concepts deployed by Marx, Heidegger, Lacan, Althusser, Agamben, Debord, Deleuze, Descartes, Sellars, Foucault, Baudrillard, Gramsci, Guattari, and Hegel. The phrases "the inverted ladder of the law of desire" and "the law of the hidden presence of the Other" are from Lacan.

There are numerous literary references contained in the sequence of performance monologues entitled "[language as a vocation]."

The poem/monologue "[the Priest]" cites Matthew 22:14.

The poem/monologue "[the Lawgiver]" paraphrases, draws from, alludes to, or directly cites Machiavelli's *The Prince*, the Code of Hammurabi, the Declaration of Independence, the Bible, the political philosophy of Thomas Hobbes (as expressed in *The Leviathan*), the political philosophy of Giorgio Agamben (as expressed in *Homo Sacer*), the Magna Carta, the Twelve Tables of Roman Law, and an infamous quote attributed to the Sun King, Louis the XIV ("*L'État c'est moi*").

The poem/monologue "[the Censor]" cites Justice John Paul Steven's dissenting opinion in the Supreme Court Case FCC v. Fox Television Stations, Inc., 556 U. S. 502, 529 (2009).

The poem/monologue "[the Bureaucrat]" amalgamates, appropriates, and reimagines textual excerpts from the State Leadership Accountability Act, a section of the California Government Code.

The poem "[preliminary notes on the status and position of

a figure pasted to the landscape]" utilizes and slightly alters material from Charles Olson's poem "La Preface" and is used by permission.

The poem "*[Untitled (1999)]*" uses Einstein's famous phrase *spooky action at a distance* to meditate on a sculpture by Anish Kapoor and references conceptual content from the placard accompanying the sculpture when the author encountered *Untitled (1999)* in the Denver Art Museum in September 2019.

The poem "[the Ear poem] can be read as meditation on the recursive nature of many forms of language and finds influence in the scholarly works of Michael C. Corbalis. For an alternative view on language and recursion see the works of Daniel L. Everett.

Additional references—to Heidegger, Beckett, and Francisco de Goya—are noted in the series of performance monologues that comprise "[meta and qua]." These poem/monologues also allude to the performance art of Andrea Fraser. The series is inspired by the poetry, drama, and hybrid-texts of Leslie Scalapino and the experimental plays and theatrical productions of Robert Wilson. Portions of an earlier version of this series of performance texts appeared in *The American River Review* No. 30.

The poem *[language and death]* takes its title from Agamben's meditative work of the same name and attempts to recapitulate Agamben's examination and exposition of Heidegger's and Hegel's treatment of the negative moment (and movement) of language using the medium of poetry to imagine the existential mechanisms by which language comes to be what it is.

The poem "[the strange appearance of the very same sentence]" paraphrases Acts 17:28 and appropriates a line from Shakespeare's *Romeo and Juliet*.

The sigil/glyph/ideogram at the end of the book was conceptually designed by the author and rendered by C.J. Rosenthal. The sigil alludes to recurring themes and concepts in philosophy and metaphysics, including the concepts "ground", "frame", and "horizon". The symbol may also be thought of as a picture for the conceptual content of this book.

3. With Gratitude

Several people provided feedback on the manuscript as it developed. No one gave more in this regard than Amy Wilson. Amy thank you for the multiple readings, all the editorial comments, and more importantly, for being such a good friend. An ear is an ear. It presupposes other ears.

Other readers to whom I am indebted include Andrew Joron, Dean Rader, Brad Buchanan, Phyllis Jeffrey, Lisa Connelly, Gabrielle Myers, David Jackson, Jodi Angel, and Sunny Jeanne Givens. Thank you.

To my sons, Jacob, Samuel, Finnegan, appreciation for your dedication to achievement and for filling your parents' lives with meaning. Sandy, thank you for being a solid co-parent to our three sons. To my parents, Dan and Georgia, and my siblings, Rebekah and Joseph, thank you for the love, support, and encouragement over the years.

Michael and Christian thank you for making this book a reality and for pushing me to sharpen the content. Christian and Paul thank you for your efforts on the design and layout.

Works Consulted,
Source Material,
Inspiration

Agamben, Giorgio. *Creation and Anarchy: The Work of Art and the Religion of Capitalism*, trans. Adam Kotsko. Stanford: Stanford University Press, 2019.

Agamben, Giorgio. *The Open: Man and Animal*, trans. Kevin Attell. Stanford: Stanford University Press, 2002.

Agamben, Giorgio. *Language and Death: The Place of Negativity*, trans. Karen E. Pinkus with Michael Hardt. Minneapolis: University of Minnesota Press, 1991.

Agamben, Giorgio. *Homo Sacer. Sovereign Power and Bare Life*. Stanford: Stanford University Press, 1998.

Ayala, Francisco J. and Camilo J. Cela-Conde, eds. *Processes In Human Evolution: The journey from early hominins to Neanderthals and modern humans*. 2nd ed. Oxford: Oxford University Press, 2017.

Badiou, Alain. *The Age of the Poets: And Other Writings on Twentieth-Century Poetry and Prose*, trans. Bruno Bosteels. London: Verso, 2014.

Bahn, Paul G. *Images of the Ice Age*, 3rd ed. Oxford: Oxford University Press, 2016.

Berardi, Franco "Bifo". *Breathing: Chaos and Poetry*. South Pasadena: Semiotext(e), 2018.

Berardi, Franco "Bifo". *The Uprising: On Poetry and Finance*. Los Angeles: Semiotext(e), 2012.

Blanchot, Maurice, George Quasha, Lydia Davis, Paul Auster, Robert Lamberton, Christopher Fynsk, and Charles Stein. *The Station Hill Blanchot Reader: Fiction & Literary Essays*. Barrytown, N.Y.: Station Hill/Barrytown, Ltd, 1999.

Brassier, Ray. *Nihil Unbound: Enlightenment and Extinction*. New York: Palgrave Macmillan, 2007.

Buchanan, Ian. *Deleuzism: A Metacommentary*. Durham: Duke University Press, 2000.

Bryant, Levi R. *Difference and Givenness: Deleuze's Transcendental Empiricism and the Ontology of Immanence*. Evanston: Northwest University Press, 2008.

Chiesa, Lorenzo. *Subjectivity and Otherness: A Philosophical Reading of Lacan*. Cambridge: MIT Press, 2007.

Colebrook, Claire. *Understanding Deleuze*. Crows Nest: Allen & Unwin, 2002.

Corballis, Michael C. *The Recursive Mind: The Origins of Human Language, Thought, and Civilization*. Princeton: Princeton University Press, 2011.

Corballis, Michael C. *The Truth About Language: What It Is and Where It Came From*. Chicago: The University of Chicago Press, 2017.

Davis, Bret W., Brian Schroeder, and Jason M. Wirth, eds. *Japanese and Continental Philosophy: Conversations with the Kyoto School*. Bloomington: Indiana University Press, 2011.

Debord, Guy. *Society of the Spectacle*. Detroit: Black and Red, 1983.

de la Durantaye, Leland. *Giorgio Agamben: A Critical Introduction*. Stanford: Stanford University Press, 2009.

Deleuze, Gilles. *Difference and Repetition*, trans. Paul Patton. New York: Columbia University Press, 1994.

Deleuze, Gilles, and Felix Guattari. *Anti-Oedipus: Capitalism and Schizophrenia*, trans. Brian Massumi. Minneapolis: University of Minnesota Press, 1983.

Deleuze, Gilles, and Felix Guattari. *Nomadology: The War Machine*, trans. Brian Massumi. New York: Semiotext(e), 1986.

Deleuze, Gilles, and Felix Guattari. *On the Line*, trans. John Johnston. New York: Semiotext(e), 1983.

Deleuze, Gilles, and Felix Guattari. *A Thousand Plateaus: Capitalism and Schizophrenia*, trans. Brian Massumi. Minneapolis: University of Minnesota Press, 1987.

Diringer, David. *The Book Before Printing: Ancient, Medieval, and Oriental*. New York: Dover, 1982.

Dreyfus, Hubert L. *Being in the World: A commentary on Heidgger's Being and Time, Division 1*. Cambridge: MIT Press, 1991.

Dufourmantelle, Anne. *In Praise of Risk*, trans. Steven Miller. New York: Fordham University Press, 2019.

Everett, Daniel L. *How Language Began: The Story of Humanity's Greatest Invention*. New York: Liveright Publishing Corporation, 2017.

Grant, Iain Hamilton. *Philosophies of Nature After Schelling*. London: Continuum, 2006.

Guthrie, R. Dale. *The Nature of Paleolithic Art*. Chicago: The University of Chicago Press, 2005.

Hands, John. *Cosmo Sapiens: Human Evolution from the Origin of the Universe*. New York: Overlook Duckworth, 2016.

Harari, Yuval N. *Sapiens: a Brief History of Humankind*. New York: Harper, 2015.

Hardt, Michael. *Gilles Deleuze: An Apprenticeship in Philosophy*. Minneapolis, University of Minnesota, 1993.

Harman, Graham. *Heidegger Explained: From Phenomenon to Thing.* Chicago: Open Court, 2007.

Heidegger, Martin. *Being and Time,* trans. John Macquarrie and Edward Robinson. New York: Harper and Row, 1962.

Heidegger, Martin. *Poetry, Language, Thought,* trans. Albert Hofstadter. New York: Perennial Classics, 2001.

Heisig, James W. *Philosophers of Nothingness: An Essay on the Kyoto School.* Honolulu: University of Hawaii Press, 2001.

Joris, Pierre. *A Nomad Poetics: Essays.* Middletown: Wesleyan University Press, 2003.

Kristeva, Julia. *Desire in Language: A Semiotic Approach to Literature and Art,* ed. Leon S. Roudiez. New York: Columbia University Press, 1980.

Kristeva, Julia. *Revolution in Poetic Language,* trans. Margaret Waller. New York: Columbia University Press, 1984.

Lacan, Jacques. *Ecrits: The First Complete Edition in English,* trans. Bruce Fink. New York: W.W. Norton, 2006.

Lewis-Williams, David and David Pearce. *Inside the Neolithic Mind.* London: thames and Hudson, 2005.

Lurker, Manfred. *The Gods and Symbols of Ancient Egypt.* New York, Thames and Hudson, 1980.

Martin, Henri-Jean. *The History and Power of Writing*. Chicago: University of Chicago Press, 1988.

Meillassoux, Quentin. *After Finitude: An Essay on the Necessity of Contingency,* trans. Ray Brassier. London: Bloomsbury Academic, 2008.

Mohen, Jean-Pierre. *Prehistoric Art: The Mythical Birth of Humanity*. Paris: Terrail, 2002.

Nishitani, Keiji. *Religion and Nothingness,* trans. Jan Van Bragt. Berkeley: University of California Press, 1982.

Nishitani, Keiji. *The Self-Overcoming of Nihilism,* trans. Graham Parkes with Setsuko Aihara. Albany: SUNYPress, 1990.

Ogg, James G., Gabbi M. Ogg, and Felix M. Gradstein. *A Concise Geologic Time Scale, 2016*. Amsterdam: Elsevier, 2016.

Oliver, Kelly. *The Portable Kristeva*. New York: Columbia University Press, 2002.

Sandford, Mariellen R., ed. *Happenings and Other Acts*. London: Routledge, 1995.

Scarre, Chris, ed. *The Human Past: World Prehistory & the Development of Human Societies*. London: Thames and Hudson, 2005.

Schuster, Aaron. *The Trouble with Pleasure: Deleuze and Psychoanalysis*. Cambridge: MIT Press, 2016.

Somers-Hall, Henry. *Deleuze's Difference and Repetition: An Edinburgh Philosophical Guide*. Edinburgh: Edinburgh University Press, 2013.

Wade, Nicholas. *Before the Dawn: Recovering the Lost History of Our Ancestors*. New York: Penguin Press, 2006.

Wiitengenstein, Ludwig, and G. E. M. Anscombe. *Philosophical Investigations*. Oxford: Blackwell, 1997.

ABOUT THE AUTHOR

Daniel Rounds is a poet, activist, and applied sociologist based in Sacramento, California. After pursuing studies in social theory, political economy, and social movements at the political science and sociology graduate programs at UCLA, he left the academy to work for organized labor. During his career he has worked for various environmental, social justice, and labor organizations as well as in the legislative and executive branches of California state government. His previous books *eros zero* (2017) and *some distant lateral present* (2014) were also published by Ad Lumen Press.